THE
COMPLETE IDIOT'S GUIDE® TO

Algebra

Second Edition

by W. Michael Kelley

ALPHA

A member of Penguin Group (USA) Inc.

For my wife, Lisa, who makes my life worth living, and my son, Nicholas, who taught me that waking up in the morning with the people you love is just the best thing in the world.

ALPHA BOOKS

Published by the Penguin Group

Penguin Group (USA) Inc., 375 Hudson Street, New York, New York 10014, USA

Penguin Group (Canada), 90 Eglinton Avenue East, Suite 700, Toronto, Ontario M4P 2Y3, Canada (a division of Pearson Penguin Canada Inc.)

Penguin Books Ltd, 80 Strand, London WC2R 0RL, England

Penguin Ireland, 25 St Stephen's Green, Dublin 2, Ireland (a division of Penguin Books Ltd.)

Penguin Group (Australia), 250 Camberwell Road, Camberwell, Victoria 3124, Australia (a division of Pearson Australia Group Pty. Ltd.)

Penguin Books India Pvt. Ltd., 11 Community Centre, Panchsheel Park, New Delhi—110 017, India

Penguin Group (NZ), 67 Apollo Drive, Rosedale, North Shore, Auckland 1311, New Zealand (a division of Pearson New Zealand Ltd.)

Penguin Books (South Africa) (Pty.) Ltd, 24 Sturdee Avenue, Rosebank, Johannesburg 2196, South Africa

Penguin Books Ltd., Registered Offices: 80 Strand, London WC2R 0RL, England

Copyright © 2007 by W. Michael Kelley

International Standard Book Number: 978-159257-648-7
Library of Congress Catalog Card Number: 2007922826

09 08 07   8 7 6 5 4 3 2 1

Interpretation of the printing code: The rightmost number of the first series of numbers is the year of the book's printing; the rightmost number of the second series of numbers is the number of the book's printing. For example, a printing code of 07-1 shows that the first printing occurred in 2007.

Printed in the United States of America

Note: This publication contains the opinions and ideas of its author. It is intended to provide helpful and informative material on the subject matter covered. It is sold with the understanding that the author and publisher are not engaged in rendering professional services in the book. If the reader requires personal assistance or advice, a competent professional should be consulted.

The author and publisher specifically disclaim any responsibility for any liability, loss, or risk, personal or otherwise, which is incurred as a consequence, directly or indirectly, of the use and application of any of the contents of this book.

Most Alpha books are available at special quantity discounts for bulk purchases for sales promotions, premiums, fundraising, or educational use. Special books, or book excerpts, can also be created to fit specific needs.

For details, write: Special Markets, Alpha Books, 375 Hudson Street, New York, NY 10014.

Publisher: *Marie Butler-Knight*
Product Manager: *Phil Kitchel*
Managing Editor: *Billy Fields*
Editorial Director/Acquiring Editor: *Mike Sanders*
Development/Copy Editor: *Ginny Bess Munroe*
Production Editor: *Kayla Dugger*

Cartoonist: *Shannon Wheeler*
Cover Designer: *Bill Thomas*
Book Designer: *Trina Wurst*
Indexer: *Angie Bess*
Layout: *Rebecca Harmon*
Proofreader: *Aaron Black*

Contents at a Glance

Contents

Introduction

Picture this scene in your mind. I am a high school student, chock-full of hormones and sugary snack cakes, thanks to puberty and the fact that I just spent the $3 my mom gave me for a healthy lunch on Twinkies and doughnuts in the cafeteria. I am young enough that I still like school, but old enough to understand that I'm not supposed to act like it, and my mind is active, alert, and tuned in. There are only two more classes to go and my day is over, and with that in mind, I head for algebra class.

In retrospect, I think the teacher must have had some sort of diabolical fun-sucking and joy-destroying laser ray gun hidden in the drop-down ceiling of that classroom, because just walking into algebra class put me in a bad mood. It's as hot as a varsity football player's armpit in that windowless, dank dungeon, and strangely enough, it always smells like a roomful of people just finished jogging in place. Vague yet acrid sweat and body odor attack my senses, and I slink down into my chair.

"I have to stay awake today," I tell myself. "I am on the brink of getting hopelessly lost, so if I drift off again, I won't understand anything, and we have a big test in a few days." However, no matter how I chide and cajole myself into paying attention, it is utterly impossible.

The teacher walks in and turns on a small oscillating fan in a vain effort to move the stinky air around and revive her class. Immediately she begins, in a soft, soothing voice, and the world in my peripheral vision begins to blur. Uh oh, soft monotonous vocal delivery, the droning white noise of a fan, the compelling malodorous warmth that only occupies rooms built out of brightly painted cinderblock … all elements that have thwarted my efforts to stay awake in class before.

I look around the room, and within 10 minutes most of the students are asleep. The few that are still conscious are writing notes to boyfriends or girlfriends. The school's star soccer player sits next to me, eyes wide and staring at his Trapper Keeper notebook, having regressed into a vegetative state as soon as class began. I begin to chant my daily mantra to myself, "I hate this class, I hate this class, I hate this class …" and I really mean it. As far as I am concerned, algebra is the most boring thing that was ever created, and it exists solely to destroy my happiness.

Can you relate to that story? Even though the individual details may not match your experience, did you have a similar mantra? Some people have a hard time believing that a math major really hated math during his formative years. I guess the math after algebra got more interesting, or my attention span widened a little bit. However, that's not the normal course of events. Luckily, my extremely bad experience with math didn't prevent me from taking more classes, and eventually my opinion changed, but most people hit the brick wall of algebra and give up on math forever in hopeless despair.

That was when I decided to go back and revisit the horribly boring and difficult mathematics classes I took, and write books that would not only explain things more clearly, but make a point of speaking in everyday language. Besides, I have always thought learning was much more fun when you could laugh along the way, but that's not necessarily the opinion of most math people. In fact, one of the mathematicians who reviewed my book *The Complete Idiot's Guide to Calculus* before it was released told me, "I don't think your jokes are appropriate. Math books shouldn't contain humor, because the math inside is already fun enough."

I believe that logic is insane. In this book, I've tried to present algebra in an interesting and relevant way, and attempted to make you smile a few times in spite of the pain. I didn't want to write a boring textbook, but at the same time, I didn't want to write an algebra joke book so ridiculously crammed with corny jokes that it insults your intelligence.

I also tried to include as much practice as humanly possible without making this book a million pages long. (Such books are hard to carry and tend to cost too much; besides, you wouldn't believe how expensive the shipping costs are if you buy them online!) Each section contains fully explained examples and practice problems to try on your own in little sidebars labeled "You've Got Problems." Additionally, Chapter 20 is jam-packed with practice problems based on the examples throughout the book, to help you identify your weaknesses if you've taken algebra before, or to test your overall knowledge once you've worked your way through the book. Remember, it doesn't hurt to go back to your algebra textbook and work out even more problems to hone your skills once you've exhausted the practice problems in this book, because repetition and practice transforms novices into experts.

Algebra is not something that can only be understood by a few select people. You can understand it and excel in your algebra class. Think of this book as a personal tutor, available to you 24 hours a day, 7 days a week, always ready to explain the mysteries of math to you, even when the going gets rough.

How This Book Is Organized

This book is presented in seven sections:

In **Part 1, "A Final Farewell to Numbers,"** you'll firm up all of your basic arithmetic skills to make sure they are finely tuned and ready to face the challenges of algebra. You'll calculate greatest common factors and least common multiples, review exponential rules, tour the major algebraic properties, and explore the correct order of operations.

In **Part 2, "Equations and Inequalities,"** the preparation is over, and it's time for full-blown algebra. You'll solve equations, draw graphs, create equations of lines, and investigate inequality statements with one and two variables.

In **Part 3, "Systems of Equations and Matrix Algebra,"** you'll find the shared solutions of multiple equations and learn the basics of matrix algebra, a comparatively new branch of algebra that's really caught on since the dawn of the computer age.

Things get a little more intense in **Part 4, "Now You're Playing with (Exponential) Power!"** because the exponents are no longer content to stay small. You'll learn to cope with polynomials and radicals, and how to solve equations that contain variables raised to the second, third, and fourth powers.

Part 5, "The Function Junction," introduces you to the mathematical function, which takes center stage as you advance in your mathematical career. You'll learn how to calculate a function's domain and range, find its inverse, and graph it without having to resort to a monotonous and repetitive table of values.

Fractions are back in the spotlight in **Part 6, "Please, Be Rational!"** You'll learn how to do all the things you used to do with simple fractions (like add, subtract, multiply, and divide them) when the contents of the fractions get more complicated.

Finally, in **Part 7, "Wrapping Things Up,"** you'll face algebra's playground bully, the word problem. However, once you learn a few approaches for attacking word problems head on, you won't fear them anymore. You'll also get a chance to practice all of your skills in the "Final Exam"; don't worry, it won't be graded.

Things to Help You Out Along the Way

As a teacher, I constantly found myself going off on tangents—everything I mentioned reminded me of something else. These peripheral snippets are captured in this book as well. Here's a guide to the different sidebars you'll see peppering the pages that follow.

You've Got Problems

Math is not a spectator sport! Once I introduce a topic, I'll explain how to work out a certain type of problem, and then you have to try it on your own. These problems will be very similar to those that I walk you through in the chapters, but now it's your turn to shine. You'll find all the answers, explained step-by-step, in Appendix A.

Kelley's Cautions

Although I will warn you about common pitfalls and dangers throughout the book, the dangers in these boxes deserve special attention. Think of these as skulls and crossbones painted on little signs that stand along your path. Heeding these cautions can sometimes save you hours of frustration.

Critical Point

These notes, tips, and thoughts will assist, teach, and entertain. They add a little something to the topic at hand, whether it be some sound advice, a bit of wisdom, or just something to lighten the mood a bit.

Talk the Talk

Algebra is chock-full of crazy- and nerdy-sounding **words** and **phrases**. In order to become King or Queen Math Nerd, you'll have to know what they mean!

How'd They Do That?

All too often, algebraic formulas appear like magic, or you just do something because your teacher told you to. If you've ever wondered "Why does that work?" or "Where did that come from?" or "How did that happen?" this is where you'll find the answer.

Acknowledgments

If I have learned anything in the short time I've spent as an author, it's that authors are insecure people, needing constant attention and support from friends, family members, and folks from the publishing house, and I lucked out on all counts. Special thanks are extended to my greatest supporter, Lisa, who never growled when I trudged into my basement and dove into my work, day in and day out (and still didn't mind that I watched football all weekend long—honestly, she must be the world's greatest wife). Also, thanks to my extended family and friends, especially Dave, Chris, Matt, and Rob, who never acted like they were tired of hearing every boring detail about the book as I was writing.

Thanks go to my agent, Jessica Faust at Bookends, LLC, who pushed and pushed to get me two great book-writing opportunities, and Nancy Lewis, my development editor, who is eager and willing to put out the little fires I always end up setting every day. Also, I have to thank Mike Sanders at Pearson/Penguin, who must have tons of experience with neurotic writers, because he's always so nice to me.

Sue Strickland, my mentor and one-time college instructor, has once again agreed to technically review this book, and I am indebted to her for her direction and expertise. Her love of her students is contagious, and it couldn't help but rub off on me.

Here and there throughout this book, you'll find in-chapter illustrations by Chris Sarampote, a longtime friend and a magnificent artist. Thanks, Chris, for your amazing drawings, and your patience when I'd call in the middle of the night and say "I think the arrow in the football picture might be too curvy."

Finally, I need to thank Daniel Brown, my high school English teacher, who one day pulled me aside and said "One day, you will write math books for people such as I, who approach math with great fear and trepidation." His encouragement, professionalism, and knowledge are most of the reason that his prophecy has come true.

Special Thanks to the Technical Reviewer

The Complete Idiot's Guide to Algebra was reviewed by an expert who double-checked the accuracy of what you'll learn here, to help us ensure that this book accurately communicates everything you need to know about algebra. Special thanks are extended to Susan Strickland, who also provided the same service for *The Complete Idiot's Guide to Calculus* (among many other titles written by me).

Susan Strickland received a Bachelor's degree in mathematics from St. Mary's College of Maryland in 1979, a Master's degree in mathematics from Lehigh University in 1982, and took graduate courses in mathematics and mathematics education at The American University in Washington, D.C., from 1989 through 1991. She was an assistant professor of mathematics and supervised student teachers in secondary mathematics at St. Mary's College of Maryland from 1983 through 2001. It was during that time that she had the pleasure of teaching Michael Kelley and supervising his student teacher experience. Since 2001, she has been a professor of mathematics at the College of Southern Maryland and is now involved with teaching math to future elementary school teachers. Her interests include teaching mathematics to "math phobics," training new math teachers, and solving math games and puzzles (she can really solve the Rubik's Cube).

Trademarks

All terms mentioned in this book that are known to be or are suspected of being trademarks or service marks have been appropriately capitalized. Alpha Books and Penguin Group (USA) Inc. cannot attest to the accuracy of this information. Use of a term in this book should not be regarded as affecting the validity of any trademark or service mark.

Part 1

A Final Farewell to Numbers

When most people think of math, they think "numbers." To them, math is just a way to figure out how much they should tip their waitress. However, math is so much more than just a substitute for a laminated card in your wallet that tells you what 15 percent of the price of your dinner is. In this part, I make sure you're up to speed with numbers and have mastered all of the basic skills you will need later on.

Chapter 1

Getting Cozy with Numbers

In This Chapter

- Categorizing types of numbers
- Coping with oodles of signs
- Brushing up on prealgebra skills
- Exploring common mathematical assumptions

Most people new to algebra view it as a disgusting, creepy disease whose sole purpose is to ruin everything they've ever known about math. They understand multiplication and can even divide numbers containing decimals (as long as they can check their answers with a calculator or a nerdy friend), but algebra is an entirely different beast—it contains *letters!* Just when you feel like you've got a handle on math, suddenly all these *x*'s and *y*'s start sprouting up all over like pimples on prom night.

Before I can even begin talking about those letters (they're actually called *variables*), you've got to know a few things about those plain old numbers you've been dealing with all these years. Some of the things I discuss in this chapter will sound familiar, but most likely, some of it will also be new. In essence, this chapter is a grab bag of prealgebra skills I need to review with you; it's one last chance to get to know your old number friends better before we unceremoniously dump letters into the mix.

Classifying Number Sets

Most things can be classified in a bunch of different ways. For example, if you had a cousin named Scott, he might fall under the following categories: people in your family, your cousins, people with dark hair, and (arguably) people who could stand to brush their teeth a little more often. It would be unfair to consider only Scott's hygiene (lucky for him); that's only one classification. A broader picture is painted if you consider all of the groups he belongs to:

♦ People in your family

♦ Your cousins

♦ People with dark hair

♦ Hygienically challenged people

The same goes for numbers. Numbers fall into all kinds of categories, and just because they belong to one group does not preclude them from belonging to others as well.

Familiar Classifications

You've been at this number classification thing for some time now. In fact, the following number groups will probably ring a bell:

♦ *Even numbers:* Any number that's *evenly divisible* by 2 is an even number, such as 4, 12, and –10.

♦ *Odd numbers:* Any number that is not evenly divisible by 2 (in other words, when you divide by 2, you get a remainder) is an odd number, like 3, 9, and –25.

♦ *Positive numbers:* All numbers greater than 0 are considered positive.

♦ *Negative numbers:* All numbers less than 0 are considered negative.

Talk the Talk

If a number is **evenly divisible** by 2, then when you divide that number by 2, there will be no remainder.

- *Prime numbers:* The only two numbers that divide evenly into a prime number are the number itself and 1 (and that's no great feat, since 1 divides evenly into every number). Some examples of prime numbers are 5, 13, and 19. By the way, 1 is not considered a prime number, due to the technicality that it's only divisible by one thing, while all the other prime numbers are divisible by two things.

- *Composite numbers:* If a number is divisible by things other than itself and 1, then it is called a composite number, and those things that divide evenly into the number (leaving behind no remainder) are called its *factors*. Some examples of composite numbers are 4, 12, and 30.

Critical Point

Technically, 0 is divisible by 2, so it is considered even. However, 0 is not positive, nor is it negative—it's just sort of hanging out there in mathematical purgatory and can be classified as both *nonpositive* and *nonnegative*.

Talk the Talk

A **factor** is a number that divides evenly into a number and leaves behind no remainder. For example, the factors of 30 are 1, 2, 3, 5, 6, 10, 15, and 30.

I don't mean to insult your intelligence by reviewing these simple categories. Instead, I mean to instill a little confidence before I start discussing the slightly more complicated classifications.

Intensely Mathematical Classifications

Math historians (if you thought regular math people were boring, you should get a load of these guys) generally agree that the earliest humans on the planet had a very simple number system that went like this: one, two, a lot. There was no need for more numbers. Lucky you—that's not true anymore. Here are the less familiar number classifications you need to understand:

- *Natural numbers:* The numbers 1, 2, 3, 4, 5, and so forth are called the natural (or *counting*) numbers. They're the numbers you were first taught as a child when you were learning to count.

- *Whole numbers:* Shove the number 0 into the natural numbers and you get the whole numbers. That's the only difference—0 is a whole number but not a natural number. (That's easy to remember; 0 looks like a *hole*, and 0 is a *whole* number.)

◆ *Integers:* Any number that has no explicit decimal or fraction is an integer. That means –4, 17, and 0 are integers, but 1.25 and $\frac{2}{5}$ are not.

◆ *Rational numbers:* If a number can be expressed as a decimal that either repeats infinitely or simply ends (called a *terminating decimal*), then the number is rational. Basically, those conditions guarantee one thing: the number is actually equivalent to a fraction, so all fractions are automatically rational. (You can remember this using the mnemonic device "Rational means fractional." The words sound roughly the same.) The fraction $\frac{1}{3}$, the terminating decimal 7.95, and the infinitely repeating decimal .8383838383… are all rational numbers.

◆ *Irrational numbers:* If a number cannot be expressed as a fraction, or its decimal representation goes on and on infinitely but the digits don't follow some obvious repeating pattern, then the number is irrational. Although many radicals (square roots, cube roots, and the like) are irrational, the most famous irrational number is π = 3.141592653589793… No matter how many thousands (or millions) of decimal places you examine, there is no pattern to the numbers. In case you're curious, there are far more irrational numbers that exist than rational numbers, even though the rationals include every conceivable fraction!

Critical Point

Because every integer is divisible by 1, each can be written as a fraction. That means every integer (take the number 3, for example) is also a rational number with 1 in the denominator (in this case $\frac{3}{1}$).

◆ *Real numbers:* If you clump all of the rational and irrational numbers together, you get the set of real numbers. Basically, any number that can be expressed as a decimal (whether it be repeating, terminating, attractive, or awkward-looking but with a nice personality) is considered a real number.

Don't be intimidated by all the different classifications. Just mark this page and check back when you need a refresher.

Example 1: Identify the categories that the number 8 belongs to.

Solution: Because there's no negative sign preceding it, 8 is a positive number. Furthermore, it has factors of 1, 2, 4, and 8 (since all those numbers divide evenly into 8), indicating that 8 is both even and composite. Additionally, 8 is a natural number, a whole number, an integer, a rational number $\left(\frac{8}{1}\right)$, and a real number (8.0).

You've Got Problems

Problem 1: Identify all the categories that the number $\frac{3}{7}$ belongs to.

Persnickety Signs

Before algebra came along, you were only expected to perform operations (such as addition or multiplication) on positive integers, but now you'll be expected to perform the same operations on negative numbers as well. The procedures you use for addition and subtraction are completely different than the ones for multiplication and division, so I discuss them separately.

Kelley's Cautions

Most textbooks write negative numbers like this: −3. However, some write the negative sign way up high like this: ⁻3. Both notational methods mean the exact same thing, although I won't use that weird, sky-high negative sign.

Addition and Subtraction

On the first day of one of my statistics courses in college, the professor asked us, "What is 5 – 9?" The answer he expected, of course, was –4. However, the first student to raise his hand answered unexpectedly. "That's impossible," he said, "You can't take 9 apples away from 5 apples—you don't have enough apples!" Keep in mind that this was a college senior, and you can begin to understand the despair felt by the professor. It's hard to learn high-level statistics when a student doesn't understand basic algebra.

Here's some advice: don't think in terms of apples, as tasty as they may be. Instead, think in terms of earning and losing money—that's something everyone can relate to, and it makes adding and subtracting positive and negative numbers a snap. If, at the end of the problem, you have money left over, your answer is positive. If you're short on cash and still owe, your answer is negative.

Example 2: Simplify 5 – (–3) – (+2) + (–7).

Solution: This is the perfect example of an absolutely evil addition and subtraction problem, but if you follow two simple steps, it becomes quite simple.

1. **Eliminate double signs (signs that are not separated by numbers).** If two consecutive signs are the same, replace them with a single positive sign. If they are different, replace them with a single negative sign.

Ignore the parentheses for a moment and work left to right. You've got two negatives right next to each other between the 5 and 3. Since those consecutive signs are the same, replace them with a positive sign. The other two pairs of consecutive signs (between the 3 and 2 and then between the 2 and 7) are different, so they get replaced by negative signs:

$$5 + 3 - 2 - 7$$

Once the double signs are eliminated, you can move on to the next step.

2. **Consider all positive numbers as money you earn and all negative numbers as money you lose to calculate the final answer.** Remember, if there is no sign immediately preceding a number, that number is assumed to be positive. (Like the 5 in this example.)

You can read the problem $5 + 3 - 2 - 7$ as "I earned five dollars, then three more, but then lost two dollars and then lost seven more." You end up with a total net loss of one dollar, so your answer is –1.

Notice that I don't describe different techniques for addition and subtraction; this is because subtraction is actually just addition in disguise—it's basically just adding negative numbers.

You've Got Problems

Problem 2: Simplify $6 + (+2) - (+5) - (-4)$.

Multiplication and Division

When multiplying and dividing positive and negative numbers, all you have to do is follow the same "double signs" rule of thumb that you used in addition and subtraction, with a slight twist. If the two numbers you're multiplying or dividing have the same sign, then the result will be positive, but if they have different signs, the result will be negative. That's all there is to it.

Example 3: Simplify the following:

(a) $5 \times (-2)$

Solution: Because the 5 and the 2 have different signs, the result will be negative. Just multiply 5 times 2 and slap a negative sign on your answer: –10.

(b) $-18 \div (-6)$

Solution: In this problem, the signs are the same, so the answer will be positive: 3.

You've Got Problems

Problem 3: Simplify the following:
(a) $-5 \times (-8)$
(b) $-20 \div 4$

Opposites and Absolute Values

There are two things you can do to a number that may or may not change its sign: calculate its opposite and calculate its absolute value. Even though these two things have similar purposes (and are often confused with one another), they work in entirely different ways.

The *opposite* of a number is indicated by a lone negative sign out in front of it. For example, the opposite of –3 would be written like this: –(–3). The value of a number's opposite is simply the number multiplied by –1. Therefore, the only difference between a number and its opposite is its sign.

$$-\left(-\frac{1}{2}\right) = \frac{1}{2} \qquad -(4) = -4$$

On the other hand, the *absolute value* of a number doesn't always have a different sign than the original number. Absolute values are indicated by thin vertical lines surrounding a number like this: $|-9|$. (You read that as "the absolute value of –9.")

Talk the Talk

The **opposite** of a number has the same value but the opposite sign of that number. The **absolute value** of a number has the same value but will always be positive.

What's the purpose of an absolute value? It always returns the positive version of whatever's inside it. Absolute value bars are sort of like "instant negative sign removers," and are so effective they should have their own infomercial on TV. ("Does your laundry have stubborn negative signs in it that just *refuse* to come out?") Therefore, $|-3|$ is equal to 3.

Notice that the absolute value of a positive number is also positive! For example, $|21| = 21$. Because absolute values only take away negative signs, if the original number isn't negative, they don't have any effect on it at all.

You've Got Problems

Problem 4: Determine the values of $-(8)$ and $|8|$.

Come Together with Grouping Symbols

The absolute value symbols I just mentioned are just one example of algebraic *grouping symbols*. Other grouping symbols include (parentheses), [brackets], and {braces}. These symbols surround portions of a math problem, and whatever appears inside the symbols is considered grouped together.

Critical Point

Technically, a fraction bar is also a grouping symbol, because it separates a fraction into two groups, the numerator and denominator. Therefore, you should simplify the two parts separately at the beginning of the problem.

Grouping symbols are important because they help you decide what to do first. Actually, there is a very specific order in which you are supposed to simplify mathematical expressions called the *order of operations*, which I discuss in greater detail in Chapter 3. Until then, just remember that anything appearing within grouping symbols should be done first.

Example 4: Simplify the following expressions.

(a) $15 \div \{7 - 2\}$

Solution: Because $7 - 2$ appears in braces, you should combine those numbers together before dividing:

$$15 \div 5 = 3$$

(b) $|5-3+(-8)|$

Solution: Because absolute value bars are present, you may be tempted to strip away all the negative signs. However, since they are inside grouping symbols, you must first simplify the expression. Eliminate double signs and combine the numbers like in Example 2.

$$|5-3-8|$$

Work left to right, subtracting 5 – 3 first.

$$=|2-8|$$
$$=|-6|$$

Now that the content of the absolute values has been completely simplified, you can take the absolute value: $|-6|=6$.

(c) $10 - [6 \times (2 + 1)]$

Solution: No grouping symbol has precedence over another. For example, you don't always do brackets before braces. However, if more than one grouping symbol appears in a problem, do the innermost set first, and work your way out.

In this problem, the parentheses are contained within another grouping symbol—the brackets—making the parentheses the innermost symbols. Therefore, you should simplify 2 + 1 first.

$$10 - [6 \times 3]$$

Only one set of grouping symbols remains, the brackets. Go ahead and simplify their contents next.

$$= 10 - 18$$

All that's left between you and the joy of a final answer is a simple subtraction problem whose answer is –8.

You've Got Problems

Problem 5: Simplify the following:
  (a) $5 \times [4 - 2]$
  (b) $|2 - (16 \div 4)|$

Important Assumptions

Just because something has a very complicated name attached to it doesn't mean that the concept is necessarily very difficult to understand. Have you ever heard of hippopotomonstrosesquippedaliophobia?

From the root word "phobia," it's obviously a fear of some kind, and based on the length and complexity of the name, you might think it's some kind of powerfully debilitating fear with an intricate neurological or psychosocial cause. Maybe it's the kind of fear that's triggered by some sort of traumatic event, like discovering that your favorite television show has been preempted again by a presidential address. (That's my greatest fear, anyway.)

Actually, hippopotomonstrosesquippedaliophobia means "the fear of long words." In my experience, whether or not they begin the class with this fear, most algebra students develop it at some point during the course. You must fight it! Although the concepts I am about to introduce have rather strange and complicated names, they represent very simple ideas. Math people, like most professionals, just give complicated names to things they think are the most important.

Talk the Talk

An algebraic property (or **axiom**) is a mathematical fact that is so obvious, it is accepted without proof.

In this case, the important concepts are algebraic properties (or *axioms*), assumptions about the ways numbers work that cannot really be verified through technical mathematical proofs, but are so obviously true that math folks (who don't usually do such rash things) assume them to be true even with no hard evidence. Even if you can't technically prove them, it's easy to demonstrate how simple and obvious they are by using real number examples (which is what I do in the following samples).

Your goal when reading about these properties is to be able to match the concept with the name, because you'll see the properties used later on in the book.

Associative Property

It's a natural tendency for people to split into social groups so they can spend more time with the people whose interests match their own. As a high school teacher, I taught kids from all the social cliques: the drama kids, the band kids, the jocks, the jerks; everyone was represented somewhere. However, no matter how they associated amongst themselves as a group, the student pool stayed the same. The same is true with numbers.

No matter how numbers choose to associate using grouping symbols, their value does not change (at least with addition and multiplication, that is). Consider the addition problem

$$(3 + 5) + 9$$

The 3 and the 5 have huddled up together, leaving the poor 9 out in the cold, wondering if it's his aftershave to blame for his role as social pariah. If you simplify this addition problem, you should start inside the parentheses, since grouping symbols always come first.

$$8 + 9 = 17$$

If I leave the numbers in the exact same order but, instead, group the 5 and 9 together, the result will be the same.

$$3 + (5 + 9)$$
$$3 + 14 = 17$$

This is called the *associative property of addition;* in essence, it means that given a string of numbers to add together, it doesn't matter which you add first—the result will be the same. As I mentioned a moment ago, there's also an *associative property for multiplication.* The answer to the multiplication problem $2 \times 6 \times 4$ doesn't change if you group the first two or the last two numbers together with parentheses.

$$(2 \times 6) \times 4 = 2 \times (6 \times 4)$$
$$12 \times 4 = 2 \times 24$$
$$48 = 48$$

 Kelley's Cautions

The operations of subtraction and division are not associative; placing grouping symbols in different spots can produce completely different results. Here's just one example that proves that division is not associative:

$$(40 \div 10) \div 2 \neq 40 \div (10 \div 2)$$
$$4 \div 2 \neq 40 \div 5$$
$$2 \neq 8$$

Commutative Property

I have a hefty commute to work—it ranges between 75 and 120 minutes one way. During my epic journeys each morning and evening, I can't help but get frustrated with inconsiderate drivers who whip in and out of lanes of traffic, just to move one or two car lengths further up the road. Even though they may get 10 or 20 feet ahead of you, a few minutes later, you usually end up passing them anyway. For all their dangerous stunt driving, they don't actually gain any ground. The moral of the story: no matter what the order of the commuters, generally, everyone gets to work at the same time. Numbers already know this to be true.

When you are adding or multiplying (once again, this property is not true for subtraction or division), the order of the numbers does not matter. Check out the multiplication problem

$$3 \times 2 \times 7$$

If you multiply left to right, $3 \times 2 = 6$, and then $6 \times 7 = 42$. Did you know that you'll still get 42 even if you scramble the order of the numbers? It's called the *commutative property of multiplication*. Need to see it in action? Here you go. (Don't forget to multiply left to right again.)

$$7 \times 3 \times 2 = 21 \times 2 = 42$$

Remember, there's also a commutative property of addition:

$$5 + 19 + 4 = 19 + 4 + 5$$
$$24 + 4 = 23 + 5$$
$$28 = 28$$

Kelley's Cautions

Here's one example that demonstrates why there's no commutative property for subtraction:

$$12 - 4 - 5 \neq 4 - 12 - 5$$
$$8 - 5 \neq -8 - 5$$
$$3 \neq -13$$

Identity Properties

Both addition and multiplication (poor subtraction and division—nothing works for them) have numbers called *identity elements*, whose job is (believe it or not) to leave numbers alone. That's right—their entire job is to make sure the number you start with doesn't change its identity by the time the problem's over.

The identity element for addition (called the *additive identity*) is 0, because if you add 0 to any number, you get what you started with.

$$3+0=3 \qquad \frac{1}{2}+0=\frac{1}{2} \qquad -37+0=-37$$

Pretty simple, eh? Can you guess what the multiplicative identity is? What is the only thing that, if multiplied by any number, will return the original number? The answer is 1; anything times 1 equals itself.

$$9\times1=9 \qquad 4\times1=4 \qquad -10\times1=-10$$

These identity elements are used in the inverse properties as well.

Inverse Properties

The purposes of the inverse properties are to "cancel out" a number to get a final result that is equal to the identity element of the operation in question. That sounds complicated, but once you see what it actually means, you'll see that it's pretty simple:

♦ **Additive Inverse Property:** Every number has an opposite, and adding a number to its opposite results in the additive identity element, 0:

$$2 + (-2) = 0 \qquad\qquad -7 + 7 = 0$$

♦ **Multiplicative Inverse Property:** Every number has a *reciprocal* defined as 1 divided by that number. When you multiply a number by its reciprocal, you get the multiplicative identity element, 1:

$$5\times\left(\frac{1}{5}\right)=1 \qquad -6\times\left(-\frac{1}{6}\right)=1$$

To understand the multiplicative inverse property, you need to know a thing or two about fractions. If you struggle with fractions, there's no need to panic—Chapter 2 will help you brush off the rust.

You've Got Problems

Problem 6: Name the mathematical properties that guarantee the following statements are true.

    (a) $11 + 6 = 6 + 11$
    (b) $-9 + 9 = 0$
    (c) $(1 \times 5) \times 7 = 1 \times (5 \times 7)$

The Least You Need to Know

- ◆ Numbers can be classified in many different ways based on characteristics ranging from their divisibility to whether or not you can write them as a fraction.

- ◆ The technique used to deal with positive and negative signs in addition and subtraction problems is slightly different than the technique used to deal with them in multiplication and division problems.

- ◆ You should always calculate the value of expressions inside grouping symbols first.

- ◆ Absolute value signs spit out the positive version of their contents.

- ◆ Mathematical properties are important (although unprovable) facts that describe intuitive mathematical truths.

Chapter 2

Making Friends with Fractions

In This Chapter

- ◆ Understanding what fractions are
- ◆ Writing fractions in different ways
- ◆ Simplifying fractions
- ◆ Adding, subtracting, multiplying, and dividing fractions

Few words have the innate power to terrify people like the word "fraction." It's quite a jump to go from talking about a regular number to talking about a weird number that's made up of two other numbers sewn together! Modern-day math teachers spend a lot of time introducing this concept to young students using toy blocks and educational manipulatives to physically model fractions, but some still stick to the old-fashioned method of teaching (like most of my teachers), which is to simply introduce the topic with no explanation and then make you feel stupid if you have questions or don't understand.

In this chapter, I help you review your fraction skills, and I promise not to tease you if you have to reread portions of it a few times before you catch on. Throughout the book (and especially in Chapters 17 and 18), you'll deal with very complicated fractions that contain variables, so you should refine your basic fraction skills now, while there are still just numbers inside them.

What Is a Fraction?

There are three ways to think of fractions, all equally accurate, and each one gives you a different insight into what makes a fraction tick. In essence, a fraction is:

- **A division problem frozen in time.** A fraction is just a division problem written vertically with a fraction bar instead of horizontally with a division symbol; for example, you can rewrite $5 \div 7$ as $\frac{5}{7}$. Although they look different, those two things mean the exact same thing.

  Why, then, would you use fractions? Well, it's no big surprise that the answer to $5 \div 7$ isn't a simple number, like 2. Instead, it's a pretty ugly decimal value. To save yourself the frustration of writing out a ton of decimal places and the mental anguish of looking at such an ugly monstrosity, leave the division problem frozen in time in fraction form.

- **Some portion of a whole number or set.** As long as the top number in a fraction is smaller than its bottom number, the fraction has a (probably hideous looking) decimal value less than one. "One what?" you may ask. It depends. For example, if you have 7 eggs left out of the dozen you bought on Sunday at the supermarket, you could accurately say that you have $\frac{7}{12}$ (read "seven-twelfths") of a dozen left. Likewise, because 3 teaspoons make up a tablespoon, if a recipe calls for 2 teaspoons, that amount is equal to $\frac{2}{3}$ (read "two-thirds") of a tablespoon.

  When considering a fraction as a portion of a whole set, the top number represents how many items are present, and the bottom number represents how many items make one complete set.

- **A failed marketing attempt.** In the late 1700s, the popularity of mathematics in society began to wane, so in a desperate attempt to increase the popularity of numbers, scientists "supersized" them, creating fractions that included two numbers for the price of one. It failed miserably, and mathematicians were forever shunned from polite society and forced to wear glasses held together by masking tape. By the way, this last one may not be true—I think I may have dreamed it.

By the way, the fancy mathematical name for the top part of the fraction is the *numerator*, and the bottom part is called the *denominator*. These terms are easily confused, so I have concocted a naughty way to help you remember which is which:

$$\frac{\text{NU}\textit{merator}}{\text{DE}\textit{nominator}}$$

As long as you read the first two letters of each word, from top to bottom, all of the mysteries of fractions will be laid "bare."

Talk the Talk _____

The top part of a fraction is its **numerator** and the bottom part is the **denominator**.

Ways to Write Fractions

You can find the actual decimal value of a fraction if you divide the numerator by the denominator, effectively thawing out the frozen division problem. A calculator yields the answer fastest, but if you're one of those people who insists on doing things the old-fashioned way, long division works as well.

For example, the decimal value of $\frac{7}{12}$ is equal to $7 \div 12$, which is 0.5833333.... Note that the digit 3 will repeat infinitely. Any digit or digits in a decimal that behave like this can be written with a bar over them like so: $0.58\bar{3}$. That bar means "anything under here repeats itself over and over again."

Some fractions, called *improper fractions*, have numerators that are larger than their denominators, such as $\frac{14}{5}$. Think of this fraction as a collection of elements (as I described in the previous section): you have 14 items, and it takes only 5 items to make one whole set. Therefore, you have enough for 2 sets (which would require 10) but not enough for 3 full sets (which would require 15). Therefore, the decimal value of $\frac{14}{5}$ is somewhere between 2 and 3 (but it's closer to 3 than 2). By the way, fractions whose numerators are smaller than their denominators are called *proper fractions*.

All improper fractions can be written as *mixed numbers*, which have both an integer and fraction part and make the actual value of the fraction easier to visualize. The improper fraction $\frac{14}{5}$ corresponds to the mixed number $2\frac{4}{5}$. Following is how I got that.

Talk the Talk _____

If the numerator is greater than the denominator, then you have an **improper fraction,** which can be left as is or transformed into a **mixed number,** which has both an integer and a fraction part. If the numerator is less than the denominator, it's a **proper fraction.**

1. Divide the numerator by the denominator. In this case, $14 \div 5$ divides in 2 times with 4 left over. Mathematically, you call 2 the *quotient* and 4 the *remainder*.

2. The quotient will be the integer portion of the mixed number (the big number out front). The fraction part of the mixed number will be the reminder divided by the improper fraction's original denominator.

Most teachers would rather you leave your answer as an improper fraction rather than express it as a mixed number, even though the implication of the term "improper fraction" might suggest that it is, in some way, wrong or a breach of manners to do so.

Critical Point _____

To convert a mixed number into an improper fraction, all you have to do is add the integer part to the fraction part. In other words, $2 + \frac{4}{5} = \frac{14}{5}$. If you don't know how to add fractions to integers, I explain it later in this chapter (see the section "Adding and Subtracting Fractions").

Simplifying Fractions

Did you know that fractions don't have to look the same to be equal? I wish this were true of humans as well, because then I might sometimes be confused with George Clooney. Alas, it is not true, and I am doomed to a life of comparison with a sea of other balding and soft-around-the-middle guys like me.

Because equivalent (or equal) fractions can take on a whole host of forms, most instructors require you to put a fraction in *simplified form*, which means that its numerator and denominator don't have any factors in common. Every fraction has one unique simplified form, which you can reach by dividing out those common factors.

Example 1: Simplify the fraction $\frac{24}{36}$.

Solution: Do 24 and 36 have any factors in common? Sure—they're both even for starters, so they have a common factor of 2. Divide both parts of the fraction by 2 in an attempt to simplify it, and you

Talk the Talk _____

Once there are no factors common to both the numerator and denominator left in the fraction, the fraction is said to be in **simplified form**. The process of eliminating the common factors is called simplifying or reducing the fraction.

get $\dfrac{24 \div 2}{36 \div 2} = \dfrac{12}{18}$. However, you're not done yet! This fraction can be simplified further, because both the numerator and the denominator can be evenly divided by 6; divide both by that common factor to get $\dfrac{12 \div 6}{18 \div 6} = \dfrac{2}{3}$. Because 2 and 3 share no common factors (other than the number 1, which is a factor of all numbers), you're finished.

By the way, even though it took me two steps to simplify this fraction, you could have done it in one step, if you realized that 12 was the *greatest common factor* of 24 and 36. If you divide both numbers by the greatest common factor, you simplify the fraction in one step; in this case, you immediately get the answer of $\dfrac{2}{3}$.

Talk the Talk

The largest factor two numbers have in common is called (quite predictably) the **greatest common factor**, and is abbreviated GCF.

If you're not convinced that the fractions $\dfrac{2}{3}$ and $\dfrac{24}{36}$ are just two different representations of the same value, convert them into decimals. It turns out they are both exactly equal to $0.\overline{6}$, conclusive evidence that they are equivalent!

You've Got Problems

Problem 1: Simplify the fractions and identify the greatest common factor of the numerator and denominator.

(a) $\dfrac{7}{21}$

(b) $\dfrac{24}{40}$

Locating the Least Common Denominator

Sometimes it's not useful to completely simplify fractions. In some cases (which I discuss in greater detail later in the chapter), you'll want to rewrite fractions so that they have the same denominator rather than being fully simplified.

That's not as hard as it seems. Remember, fractions (like $\dfrac{2}{3}$ and $\dfrac{24}{36}$) can look dramatically different but actually have the exact same value. The tricky part of rewriting fractions to have common (equal) denominators is figuring out exactly what that common denominator should be. However, if you follow these steps, identifying a common denominator should pose no challenge at all:

1. Examine the denominators of all the fractions. Choose the largest of the group. For grins, I will call this large denominator Bubba.

2. Do all of the other, smaller denominators divide evenly into Bubba? If so, then Bubba is your least common denominator. If not, proceed to step 3.

3. Multiply Bubba by 2. Do all of the other denominators *now* divide evenly into Bubba? If not, multiply Bubba by 3 and see if that works. If not, continue multiplying Bubba by bigger and bigger numbers until all the denominators divide evenly into that big boy.

Not only does this procedure generate a common denominator, it actually generates the smallest one, called the *least common denominator* (abbreviated LCD).

Kelley's Cautions

Some students cop out when calculating a common denominator; rather than follow my simple three-step procedure, they multiply all of the denominators together. For example, to find a common denominator for the fractions $\frac{7}{200}$, $\frac{13}{100}$, and $-\frac{8}{50}$, they would multiply 200 × 100 × 50 = 1,000,000, a gigantic common denominator!

Although that is a valid common denominator, it is certainly not the smallest one. Using my technique, 200 is the largest denominator (hence it is named Bubba), and since the other denominators (100 and 50) both divide into Bubba evenly, Bubba is the least common denominator. I don't know about you, but I'd much rather deal with a denominator of 200 than one that's 5,000 times as large!

Once you've calculated the least common denominator, you're halfway done. You still have to rewrite the fractions so that they contain their brand new, shiny common denominators. Here's how to do it:

1. Divide the least common denominator by the denominator of each fraction. (Everything will divide evenly.)

2. Multiply both the numerator and denominator of each fraction by the result.

I bet this whole process sounds intimidating. It takes a little practice until it becomes second nature to you, but (as you'll see in the next few examples) there's no one step that's overly complicated.

Example 2: Rewrite the fractions so they contain the least common denominator.

(a) $\dfrac{1}{3}, \dfrac{7}{15}$

> **Solution:** First, look for the biggest denominator (Bubba): 15. Because the other denominator (3) divides into 15 evenly, that makes 15 the least common denominator, which is handy because it means the second fraction won't need to be rewritten. However, the first fraction must be altered so that it also has a denominator of 15.
>
> To rewrite the first fraction, divide the LCD by the denominator (15 ÷ 3) and multiply the fraction's numerator and denominator by the result (5).
>
> $$\frac{1 \times 5}{3 \times 5} = \frac{5}{15}$$
>
> Notice that the new fraction is not in simplified form (if you were to simplify it, you'd get back what you started with, $\frac{1}{3}$). Your final fractions, now written with common denominators, are $\dfrac{5}{15}$ and $\dfrac{7}{15}$.

How'd They Do That?

When you multiply the numerator and denominator of a fraction by the same number, it's like you're multiplying by 1, which won't affect the fraction's value because 1 is the multiplicative identity. That means the original $\left(\frac{1}{3}\right)$ and final fractions $\left(\frac{5}{15}\right)$ have the same value.

(b) $\dfrac{1}{2}, \dfrac{1}{3},$ and $\dfrac{3}{4}$

> **Solution:** Bubba equals 4 in this set of fractions. Even though 2 divides into Bubba evenly, 3 does not. So, multiply Bubba by 2 (4 × 2) to get 8. Stubbornly, 3 *still* won't divide evenly into 8. Oh well, onwards and upwards. Now multiply Bubba by 3 (4 × 3) to get 12. All of the denominators divide evenly into 12, so 12 is the least common denominator.

To finish the problem, you have to multiply both the numerator and denominator of each fraction by the appropriate number. (Remember, that number is the result of dividing the least common denominator, 12, by the individual fraction's denominator).

$$\frac{1 \times 6}{2 \times 6} = \frac{6}{12} \qquad \frac{1 \times 4}{3 \times 4} = \frac{4}{12} \qquad \frac{3 \times 3}{4 \times 3} = \frac{9}{12}$$

You've Got Problems

Problem 2: Rewrite the following fractions using the least common denominator.
$$\frac{1}{3}, \frac{5}{6}, \frac{7}{10}$$

Operations with Fractions

Now that you have a basic knowledge about what fractions are and how you can manipulate them, it's time to start combining them using the four basic arithmetic operations.

Adding and Subtracting Fractions

In case you're wondering why I spent so much time discussing common denominators, you're about to find out. (Isn't this exciting?) It turns out that *you can only add or subtract fractions if they have common denominators.* If I had a quarter for every time a student of mine saw the problem $\frac{2}{3} + \frac{9}{11}$ and added the numerators and denominators together to get an answer of $\frac{11}{14}$, I might not be a rich man, but I could spend most of a three-day weekend at the mall playing pinball.

Here's the correct way to add or subtract fractions:

1. Write the fractions using a common denominator. (If adding a fraction to an integer, write the integer as a fraction by dividing it by 1 before getting common denominators.)

2. Add the numerators of the fractions and write the result over the common denominator.

3. Simplify the fraction if necessary.

The hardest part of the entire process is getting that common denominator, and since you already know how to do that, this fraction thing is no sweat!

Example 3: Combine the fractions as indicated and express the answer in simplest form.

(a) $\dfrac{3}{4} - \dfrac{8}{3}$

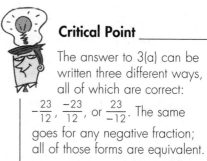

Critical Point

The answer to 3(a) can be written three different ways, all of which are correct: $-\dfrac{23}{12}$, $\dfrac{-23}{12}$, or $\dfrac{23}{-12}$. The same goes for any negative fraction; all of those forms are equivalent.

    Solution: Even though one of these fractions is improper, you don't do anything differently. Start by rewriting the fractions with the least common denominator 12.

$$\frac{3\times3}{4\times3} - \frac{8\times4}{3\times4} = \frac{9}{12} - \frac{32}{12}$$

Subtract the numerators and write the result over the common denominator.

$$\frac{9-32}{12} = \frac{-23}{12}$$

Because 23 is a prime number, it won't have any factors in common with 12, so there's no way to simplify the fraction.

(b) $2 + \dfrac{4}{5} + \dfrac{7}{10}$

    Solution: Start by rewriting 2 as a fraction. (All integers have an invisible denominator of 1.)

$$\frac{2}{1} + \frac{4}{5} + \frac{7}{10}$$

Rewrite the fractions using the least common denominator, 10.

$$= \frac{2\times10}{1\times10} + \frac{4\times2}{5\times2} + \frac{7}{10}$$

$$= \frac{20}{10} + \frac{8}{10} + \frac{7}{10}$$

Add the numerators together and divide the answer by the common denominator.

$$= \frac{20+8+7}{10}$$

$$= \frac{35}{10}$$

Simplify the fraction by dividing both numbers by the greatest common factor, 5.

$$= \frac{7}{2}$$

You've Got Problems

Problem 3: Combine the fractions and express the answer in simplest form.

$$\frac{3}{2} - \frac{1}{3} + \frac{1}{12}$$

Multiplying Fractions

While it's a shame you can't add two fractions by just summing up their respective numerators and denominators, it's a nice surprise that multiplying fractions is just that easy. You read right! All you have to do is multiply all of the numerators together and write the result over all of the denominators multiplied together.

Example 4: Multiply the fractions and express the answer in simplest form.

$$\frac{1}{2} \times \frac{3}{5} \times \frac{8}{9}$$

Solution: Multiply the numerators, and then do the same for the denominators.

$$1 \times 3 \times 8 = 24 \qquad 2 \times 5 \times 9 = 90$$

The answer is the numerator product divided by the denominator product. Make sure to simplify using the greatest common factor of 6.

$$\frac{24}{90} = \frac{24 \div 6}{90 \div 6} = \frac{4}{15}$$

You've Got Problems

Problem 4: Multiply and express the answer in simplest form.

$$\frac{5}{6} \times 8 \times \frac{3}{20}$$

Dividing Fractions

In Chapter 1, you learned that every number has a reciprocal. At that time, I told you the reciprocal was just 1 divided by the number. For example, the reciprocal of 8 is $\frac{1}{8}$.

What if, however, you want to take the reciprocal of a fraction? Is the reciprocal of $\frac{3}{5}$ equal to $\frac{1}{3/5}$? That hideous-looking thing is called a *complex fraction*, and I'll talk more about that way ahead in Chapter 17.

For now, this is what you should remember: *The reciprocal of a fraction is formed by reversing the parts of the fraction* (the numerator becomes the denominator and vice versa). For example, the reciprocal of $\frac{3}{5}$ is equal to $\frac{5}{3}$, the fraction flipped upside down.

(It's easy to remember what the word "reciprocal" means; just remember "re*flip*rocal.")

Wonder why I'm dredging up reciprocals again? Here's why: dividing by a fraction is exactly the same thing as multiplying by its reciprocal. Therefore, dividing a number by $\frac{3}{5}$ is the same as multiplying it by $\frac{5}{3}$.

Example 5: Divide the fractions and express the result in simplest form.

$$-\frac{3}{4} \div \frac{5}{16}$$

Solution: Don't be psyched out by the negative sign; even though they're fractions, they're still just numbers, and the rules you used in Chapter 1 still apply. (Because the fractions have different signs, the result will be negative.) Meanwhile, to solve the problem, take the reciprocal of the number you're dividing by and change the division sign to multiplication.

$$-\frac{3}{4} \times \frac{16}{5}$$

All that's left is a measly old multiplication problem, and you already know how to do that.

$$= -\frac{3 \times 16}{4 \times 5} = -\frac{48}{20} = -\frac{48 \div 4}{20 \div 4} = -\frac{12}{5}$$

Before I wrap up this chapter on fractions, let me hit you with an example problem that will truly see if you have mastered fractions or not. It'll test a couple of your skills at once.

You've Got Problems

Problem 5: Write the result as a fraction in simplest form.

$$\left(\frac{1}{2} - \frac{2}{7}\right) \div \frac{1}{7}$$

The Least You Need to Know

- The top of the fraction is called the numerator and the bottom is called the denominator.

- If a fraction is in simplest form, its numerator and denominator have no common factors.

- In order to add or subtract fractions, they must all have a common denominator.

- Reverse the numerator and denominator of a fraction to create its reciprocal.

- Dividing by a fraction is the same as multiplying by its reciprocal.

Chapter

3

Encountering Expressions

In This Chapter

- ◆ Representing numbers with variables
- ◆ Simplifying complex mathematical expressions
- ◆ Unleashing the power of exponents
- ◆ Performing operations in the correct order

The fateful hour has come. Tension hangs thickly in the air like a big, fat seagull. The world, as one, holds its breath as its population understands, though only subconsciously, that something important is about to happen. Little children look toward the horizon expectantly, trying to catch a glimpse of what is to come. Dogs bark without cease, straining at their leashes to run, catching the scent of revolution in the air. Nerds everywhere stop memorizing lines from *Monty Python* movies for a moment, sensing mathematical evolution.

How's that for dramatic buildup? Perhaps it's a little overstated, but everything you used to know about math is going to change, so I thought it fitting. From this point forward, every chapter is going to contain a little less of the familiar, and will be further removed from the friendly world of numbers and arithmetic. It's time to get algebratized.

Take a deep breath and wade with me out into the cool, but shallow, waters of algebraic expressions. We spend a little time in knee-deep mathematical waters during this chapter, but before long you'll be swimming in water too deep to stand up in, and you'll wonder why you thought you'd never be able to do it.

Introducing Variables

A *variable* is a letter that represents a number; it works sort of like a pronoun does in English. Rather than saying "Dave is a funny guy, because Dave can do an uncanny impression of Dave's grandmother," it's much more natural to say "Dave is a funny guy because *he* can do an uncanny impression of *his* grandmother." The pronouns *he* and *his* both clearly refer to Dave so there's no confusion on the part of the reader, and the second sentence sounds a whole lot more natural.

Remember the technique I outlined for calculating the least common denominator in Chapter 2? I called the largest original denominator *Bubba*, which is actually a variable (although variables are usually letters, not Southern-sounding nicknames). Once I defined *Bubba*, I could have shortened my explanation using the full power of variables like so: "I will call the largest denominator *Bubba* and the other denominators *a* and *b* (it does not matter which is which). If both *a* and *b* divide evenly into *Bubba*, then *Bubba* is your least common denominator. If not, calculate 2 × *Bubba*, and see if *a* and *b* divide evenly into that. If they don't, try 3 × *Bubba*, then 4 × *Bubba*, and so on until you find a number into which *a* and *b* divide evenly."

See how much easier it is to say "2 × *Bubba*" rather than "two times the largest denominator—the one you identified in the previous step"? Besides simplifying things, variables serve another important purpose—they allow you to speak very precisely. Precision is important to mathematicians, who want no confusion whatsoever in their explanations.

Critical Point

In this book (like in all math textbooks) variables are italicized so that they stand out from the surrounding words and numbers.

In Chapter 2, I also described a fraction's reciprocal as the fraction flipped upside down. Variables make it possible to create a more precise explanation: "The reciprocal of the fraction $\frac{a}{b}$ (such that *a* and *b* are integers) is the fraction $\frac{b}{a}$." There's no need for me to say "the old numerator becomes the new denominator," because you can see the old numerator, *a*, actually become the new denominator.

As useful as variables are, some math instructors overuse them. While they may provide shorter and more concise definitions, they are also more confusing to understand, especially for new algebra students. Therefore, I will almost always accompany formulas, variables, and definitions with a plain English explanation, so that you can understand what all that dense math language means.

Translating Words into Math

It's no big surprise that variables act like pronouns, because (believe it or not) math is basically its own language, defined by numeric and logical rules. Therefore, one of the first skills you must master is translating English into this new (much geekier) math language. For now, you'll translate English phrases into mathematical phrases, called *expressions.*

The expressions you'll translate are very straightforward. For example, the phrase "3 more than an unknown number" becomes the mathematical expression "$x + 3$." Because the unknown number has no explicitly stated value, you label it using the variable x. To get a value 3 more than x, simply add 3 to it. If this is tricky for you, think in terms of a real number example. Ask yourself "What is 3 more than the number 7?" Clearly the answer is 10, and you get that answer via the expression "$7 + 3 = 10$." So, to get 3 more than an unknown number, replace the known number 7 with the unknown number x.

In this translation example, the key phrase was "more than," because it clued you in that addition will be necessary. Each operation has its own cue words, and I've listed them here, with an example for each:

Kelley's Cautions

Because subtraction is not commutative, be careful to get the order of the numbers right. Notice that "5 *less* a number" and "5 less than a number" mean completely different things.

Critical Point

If no symbol is written between two things, multiplication is implied. Thus, $5y$ means "5 times y" and $2(x + 1)$ means "2 times the whole expression $(x + 1)$." Also, from this point forward, I will usually use the symbol "·" to indicate multiplication, because it's too easy to mix up the other multiplication symbol, ×, with the variable x.

Addition

♦ *More than/greater than:* "11 greater than a number" means "$x + 11$"

♦ *Sum:* "The sum of a number and 6" means "$x + 6$"

Subtraction

♦ *Fewer than/less than:* "7 fewer than a number" means "$x - 7$"

♦ *Less:* "17 less a number" means "$17 - x$"

♦ *Difference:* "The difference of a number and 6" means "$x - 6$"

Multiplication

♦ *Product:* "The product of a number and 3" means "$x \cdot 3$" or "$3x$" (because multiplication is commutative, the order in which you write the numbers doesn't matter)

♦ *Of:* "Half of 20" means "$\frac{1}{2} \cdot 20$"

Division

♦ *Quotient:* "The quotient of 10 and a number" means "$10 \div x$"

♦ *Fractions:* Any expression written as a fraction is technically a division problem in disguise

Example 1: Translate the phrases below into mathematical expressions:

(a)   The sum of 6 and twice a number

Solution: The phrase "twice a number" translates into two times the number, or $2x$. (Think about it—twice the number 8 would be $2 \cdot 8 = 16$.) The word "sum" indicates that you should add 6 and $2x$ together to get a final answer of $6 + 2x$.

(b)   The product of 2 and 3 more than a number

Solution: You may be tempted to give an answer of $2 \cdot x + 3$, but that equals $2x + 3$, or "3 more than 2 times a number." You need to use parentheses to keep the $x + 3$ together like so: $2(x + 3)$. Now, 2 is multiplied by all of $x + 3$, not just x.

You've Got Problems

Problem 1: Translate the phrase "5 less than one third of a number" into a mathematical expression.

Behold the Power of Exponents

You may have seen teeny little numbers floating above and to the right of other numbers and variables, like in the expression x^3. What is that little 3 doing up there? Parasailing? Is it small because of scale, perhaps because it is actually as large as the sun but is only seen from hundreds of thousands of miles away? No, that little guy is called the *exponent* or the *power* of x in the expression, and indeed it has a very *power*ful role in algebra.

Talk the Talk

In the exponential expression y^4, 4 is the **exponent** and y is the base.

Big Things Come in Small Packages

The role of an *exponent* is to save you time and to clean up the way expressions are written. Basically, an exponent is a shorthand way to indicate repeated multiplication. In the language of algebra, x^3 (read "x to the third power") means "x multiplied by itself three times," or $x \cdot x \cdot x$. You multiply the base (the large number) by itself as many times as the exponent indicates.

Critical Point

Two exponents have special names. Anything raised to the second power is said to be *squared* (5^2 can be read "5 squared"), and anything to the third power is said to be *cubed* (x^3 can be read "x cubed").

Example 2: Evaluate the exponential expressions:

(a) 4^3

> **Solution:** In this expression, 4 is the base and 3 is the exponent. To find the answer, multiply 4 by itself 3 times:
>
> $$4 \cdot 4 \cdot 4 = 16 \cdot 4 = 64$$
>
> Therefore, $4^3 = 64$.

(b) $(-2)^5$

> **Solution:** In this case, the base is –2, and it should be multiplied by itself 5 times.
>
> $$(-2)^5 = (-2)(-2)(-2)(-2)(-2)$$
>
> Don't stress out about the negative signs—just go left to right and multiply two numbers at a time. Start with $(-2)(-2) = 4$ and then multiply that result by the next –2, and that result by the next –2, and so on.
>
> $$(-2)(-2)(-2)(-2)(-2) = 4(-2)(-2)(-2) = -8(-2)(-2) = 16(-2) = -32$$

You've Got Problems

Problem 2: Evaluate the expression $(-3)^4$.

Exponential Rules

Once you write something in exponential form, there are very specific rules you must follow to simplify those expressions. Here are the five most important rules, each with a brief explanation:

Critical Point _____

Any number raised to the 1 power equals the original number ($x^1 = x$); so, if there's no power written, assume the power is 1 ($7 = 7^1$). In addition, anything (except 0) raised to the 0 power equals 1 ($x^0 = 1$, $12^0 = 1$). The expression 0^0 works a little differently, but you don't deal with that until calculus.

♦ **Rule 1: $x^a \cdot x^b = x^{a+b}$.** If exponential expressions with the same base are multiplied, the result is the common base raised to the *sum* of the powers.

$$x^4 \cdot x^7 = x^{4+7} = x^{11} \qquad (2^2)(2^3) = 2^{2+3} = 2^5 = 32$$

♦ **Rule 2: $\dfrac{x^a}{x^b} = x^{a-b}$.** If you are dividing exponential expressions with the same base, the result is the common base raised to the *difference* of the two powers.

$$\frac{z^7}{z^4} = z^{7-4} = z^3 \qquad \frac{(-5)^{10}}{(-5)^9} = (-5)^{10-9} = (-5)^1 = -5$$

♦ **Rule 3: $\left(x^a\right)^b = x^{a \cdot b}$.** If an exponential expression is itself raised to a power, multiply the exponents together. This is different from Rule 1, because here one base is raised to two powers, and in Rule 1 there were two bases raised to two powers.

$$\left(3^5\right)^6 = 3^{5 \cdot 6} = 3^{30} \qquad \left(k^2\right)^0 = k^{2 \cdot 0} = k^0 = 1$$

♦ **Rule 4: $\left(xy\right)^a = x^a y^a$ and $\left(\dfrac{x}{y}\right)^a = \dfrac{x^a}{y^a}$.** If a product (multiplication problem) or quotient (division problem) is raised to a power, then so is every individual piece within.

$$\left(5y\right)^2 = 5^2 \cdot y^2 = 25y^2 \qquad \left(x^2 y^3\right)^4 = \left(x^2\right)^4 \cdot \left(y^3\right)^4 = x^8 y^{12}$$

♦ **Rule 5: $x^{-a} = \dfrac{1}{x^a}$ and $\dfrac{1}{x^{-a}} = x^a$.** If something is raised to a negative power, move it to the other part of the fraction (if it's in the numerator, send it to the denominator and vice versa) and make the exponent positive. If the expression contains other positive exponents, leave them alone.

Most teachers consider answers containing negative exponents unsimplified, so make sure to eliminate negative exponents from your final answer. Also, note that raising something to the –1 power is the same as taking its reciprocal.

$$\frac{x^{-3}y^2}{z^3} = \frac{y^2}{x^3 z^3} \qquad \left(\frac{4^2}{w^5}\right)^{-1} = \frac{4^{2(-1)}}{w^{5(-1)}} = \frac{4^{-2}}{w^{-5}} = \frac{w^5}{4^2} = \frac{w^5}{16}$$

Most of the time, you'll have to apply multiple rules in the same problem (like in the next example).

Example 3: Simplify the expression $\dfrac{\left(x^2 y^{-3}\right)^2}{\left(xy^2\right)^4}$.

Solution: Start by applying Rules 3 and 4 to the numerator and denominator.

$$\frac{x^{2\cdot2}\,y^{-3\cdot2}}{x^{1\cdot4}\,y^{2\cdot4}} = \frac{x^4 y^{-6}}{x^4 y^8}$$

Now apply Rule 2, because you have matching bases in the numerator and denominator.

$$\left(x^{4-4}\right)\left(y^{-6-8}\right) = x^0 y^{-14} = 1 \cdot y^{-14} = y^{-14}$$

Finish by applying Rule 5.

$$y^{-14} = \frac{1}{y^{14}}$$

You've Got Problems

Problem 3: Simplify the expression $\left(x^3 y\right)^5 \left(x^{-2} y^2\right)^3$.

Living Large with Scientific Notation

Have you ever owned a Rubik's Cube? It's a little puzzle with six differently colored sides. According to the manufacturer, there are 43 quintillion different possible moves (all of which I have probably tried, unsuccessfully, to solve it). In case you don't know how many a quintillion is (I had to check it myself), here's the number: 43,000,000,000,000,000,000.

Numbers this large, with this many zeroes, are usually written in *scientific notation*, a method used to express numbers that are extremely large or extremely small. Here's what that number looks like in scientific form: 4.3×10^{19}. Sure it's weird-looking, but it's a heck of a lot easier to write.

In order to express a number in scientific notation, here's what you do:

♦ **Large numbers:** Move the decimal place in the original number to the left, counting each digit you pass, until only one nonzero digit remains to the left of the decimal place. Write the smaller decimal (ignoring any 0's at the end) and add this to the end of it: " $\times 10^{n}$," where n is the number of digits you passed.

Talk the Talk

You can use **scientific notation** to express numbers that are extremely large or extremely small. The end result of scientific notation is a decimal times the number 10, which is raised to a positive or negative power.

In the Rubik's example, there is no explicit decimal point, so imagine it falls at the very end of the number. You want to move it left until it passes everything but the first digit (4). You'll have to pass 19 numbers along the way, hence the notation 4.3×10^{19}.

♦ **Small numbers:** Move the decimal point to the right until exactly one nonzero digit appears to the left of the decimal point. Write the resulting decimal (without any 0s after it) and attach " $\times 10^{-n}$" to it, where n is once again the number of digits you passed along the way.

Example 4: Write the number 0.0000000000372 in scientific notation.

Solution: Because this is a very small number, move the decimal to the right until there's exactly one nonzero digit (in this case 3) to the left of the decimal point: 3.72. In order to do that, the decimal must pass 11 digits (ten 0s and the 3). Therefore, your final answer is 3.72×10^{-11}. The exponent is negative because you moved the decimal right; it's positive when the decimal moves left.

You've Got Problems

Problem 4: Write the numbers in scientific notation:
(a) 23,451,000,000,000
(b) 0.00000000125

Dastardly Distribution

Many times throughout the rest of the book, you're going to see a number or variable multiplied by a quantity in parentheses, like this: $5(x + 1)$, which is read "5 times the quantity x plus 1." The parentheses are used to indicate that the 5 is multiplied by the *entire* quantity $x + 1$, not just the x nor just the 1.

According to an algebraic property (that I kept secret until now … surprise!) called the *distributive property*, you can rewrite that expression as $5x + 5$. In other words, you can multiply the 5 times everything inside those parentheses.

I think of the x and the 1 as my brother and I at Christmas time, and the 5 as a present from my grandmother. She knew that if she got us different presents, there would be strife in the house. Both of us would love our present but secretly covet the other guy's gift as well. To avoid this, my grandmother would give us both the same thing. That's how the distributive property works—it gives the same thing (in the form of multiplication) to everything in the parentheses to avoid conflict and, possibly, brawling beneath the mistletoe.

Talk the Talk

The **distributive property** is defined mathematically like this: $a(b + c) = ab + ac$. Notice that a is multiplied by everything inside the parentheses, which are then dropped.

Critical Point

Occasionally, you'll see an expression like $2x - (3x - 5)$. Even though it's not written as such, this means the same as $2x - 1(3x - 5)$; the negative sign in front of those parentheses might as well be a -1. To simplify the expression, distribute -1:

$$2x - 1(3x) - 1(-5) = 2x - 3x + 5 = -x + 5$$

Example 5: Apply the distributive property: $-2x(3x^2 + 5y - 7)$.

Solution: Multiply everything in the parentheses by $-2x$ and add the results together.

$$(-2x)(3x^2) + (-2x)(5y) + (-2x)(-7)$$

Uh-oh … how do you multiply $-2x$ by $3x^2$? It's easy—just multiply the number parts together ($-2 \cdot 3 = -6$) and the variable parts together using exponential rules ($x \cdot x^2 = x^{1+2} = x^3$) to get $-6x^3$. Do the same thing with the remaining products.

$$-6x^3 - 10xy + 14x$$

Notice that there's no handy way to multiply x and y together in the second term. They aren't exponential expressions with the *same* base, so you can't add the powers—just leave them be and they won't hurt anyone, I promise.

You've Got Problems

Problem 5: Apply the distributive property: $6y^3(2x - 5y^2 + 8)$.

Get Your Operations in Order

If I asked you to evaluate the expression $4 + 5 \cdot 2$, what answer would you give? Most new algebra students would answer 18, but the correct answer is actually 14. Math isn't like reading—you don't always perform the operations in an expression left to right. Instead, the order depends upon the operations themselves. For example, you should always perform multiplication before addition, so in the expression $4 + 5 \cdot 2$, you should multiply $5 \cdot 2$ to get 10 and *then* add 4, which is where the correct answer of 14 comes from. Here's the exact order you should follow when evaluating expressions:

1. **Parentheses (or other grouping symbols):** If there's more than one set of grouping symbols, start with the innermost set and work your way out.

2. **Exponents:** Check to see if you can simplify anything with a power attached to it.

3. **Multiplication and Division:** These operations are done in the same step working from left to right.

4. **Addition and Subtraction:** Like multiplication and division, you should both add and subtract in the same step, again working from left to right.

You may have heard the mnemonic phrase "**P**lease **e**xcuse **m**y **d**ear **A**unt **S**ally" used to help remember the order of operations. The first letters of each word give you the correct order of operations: **p**arentheses, **e**xponents, **m**ultiplication, **d**ivision, **a**ddition, and **s**ubtraction. However, this phrase often confuses students, who see "m" come before "d" in the phrase and assume that multiplication always comes before division, which is not necessarily true—they're done in the same step from left to right. If you keep that in mind, though, it's a very handy phrase. If the Aunt Sally phrase is too outdated or stale for you, try "**P**oetic **e**lephants **m**ight **d**ream **a**bout **s**onnets" or "**P**enguins **e**at **m**y **d**ad's **a**thletic **s**ocks."

Example 6: Simplify the expressions:

(a) $1 + 10 \div (4 - 2)$

   Solution: Parentheses come first, so subtract 2 from 4.

$$1 + 10 \div 2$$

   All that's left are addition and division, and according to the order of operations, division comes first.

$$1 + 5 = 6$$

(b) $6 - [(12 + 3) \div 5 + 1]$

   Solution: There are two sets of grouping symbols, so start with the innermost set, the parentheses, first: $(12 + 3) = 15$.

$$6 - [15 \div 5 + 1]$$

   You've still got brackets left, so simplify them, making sure to do division before addition. (The order of operations still applies inside those brackets.)

$$6 - [3 + 1] = 6 - 4 = 2$$

(c) -3^2

   Solution: Let me warn you—this is kind of tricky, because most people assume that -3^2 is the same as $(-3)^2$, and it's not. The expression $(-3)^2$ translates into "−3 to the second power" or $(-3)(-3)$, which equals a positive 9.

   The other expression, -3^2, means "the opposite of 3 squared." Because $3^2 = 9$, then $-3^2 = -9$.

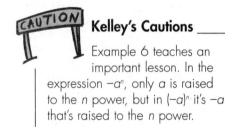

Kelley's Cautions

Example 6 teaches an important lesson. In the expression $-a^n$, only a is raised to the n power, but in $(-a)^n$ it's $-a$ that's raised to the n power.

You've Got Problems

Problem 6: Simplify the expression $10^2 \div 5^2 \cdot 3$.

Evaluating Expressions

I've used the word "evaluate" a lot so far, and as you've probably already figured out by now, it's just a fancy word for "give me the numeric answer."

Example 7: Evaluate the expression $-2x^2y$ if $x = 4$ and $y = -3$.

Solution: You're given the values of x and y, so replace (substitute) the variables with the corresponding number.

$$-2(4)^2(-3)$$

Now perform the operations in the correct order—evaluate the exponent first and then multiply from left to right.

$$-2(16)(-3) = (-32)(-3) = 96$$

You've Got Problems

Problem 7: Evaluate the expression $x(y - 3)^2$ if $x = 5$ and $y = -1$.

The Least You Need to Know

♦ Exponents are used to indicate repeated multiplication, and there are specific rules for simplifying exponential expressions.

♦ Scientific notation is a shorter way to write very large or very small numbers.

♦ The distributive property allows you to multiply something by an entire expression that's within a set of grouping symbols.

♦ You must perform algebraic operations in a specific order when simplifying an expression.

Part 2

Equations and Inequalities

Lines are nothing new to you. You've drawn them, stood in them, forgotten them on stage, and even enjoyed them on corduroy pants. In this part, however, you're going to get to know lines better than you ever have before. It'll be like getting stuck in an elevator with a neighbor you barely know. Maybe you'd be uncomfortable at first, but after awhile, you'd start exchanging pleasantries, and before long you'd learn all sorts of interesting things you would never have discovered if technology hadn't pinned the two of you together in a nightmarishly small metal box.

Chapter 4

Solving Basic Equations

In This Chapter

- ◆ Calculating solutions to equations
- ◆ Rules for manipulating equations
- ◆ Isolating variables within formulas
- ◆ Solving equations containing absolute values

If algebraic expressions are the logical equivalents of sentence fragments, then equations are the equivalents of full sentences. What's the difference between a phrase and a full sentence? The presence of a verb, and in the case of equations, the mathematical verb will be "equals."

Believe it or not, I think you'll find equations refreshing, especially compared with the stuff from the first few chapters. Not only was that review, but it had one major weakness: there was no easy way to tell if you got a given problem right! Sure, if asked to divide two fractions, you can probably remember to take a reciprocal and multiply, but if you make an arithmetic mistake, you're doomed! You might have had the right idea, and knew what you were supposed to do, but because of a simple screw-up (like multiplying 3 by 7 and getting 10), your test paper will be covered in so much red ink that it will look like it was attacked by a badger.

That's not true for equations. When you finish solving an equation, you can test your answer and instantly find out if you were right or not. If something went wrong, you can go back and fix it before it gets marked wrong, leaving the badger to do nothing but make whatever sounds badgers make when they're frustrated.

Maintaining a Balance

In case you haven't seen one, here's a basic equation:

$$3x - 2 = 19$$

It looks a lot like an expression, except it contains an equal sign. Just like expressions, equations translate easily into words. The above equation means "2 less than 3 times some number is equal to 19." Your job will be to figure out exactly what that number is. In this case, $x = 7$; you can probably figure that out by plugging in a bunch of different things for x until something works. However, there are better (and more trustworthy) ways to solve equations.

Think of an equation as a playground seesaw. The left and right sides of the equation sit on opposite sides of the seesaw, and because they are equal, it is exactly balanced, as in Figure 4.1.

Figure 4.1

Because the two sides of the equation are equal, it's as though their weights precisely balance them on a seesaw.

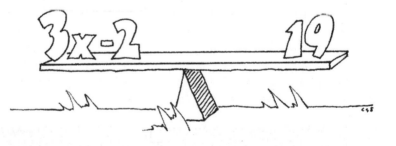

Talk the Talk

The process of forcibly isolating a variable on one side of the equal sign (usually the left side) is called **solving for** that variable.

Your job will be to shift around the contents of the seesaw so that only x remains in the left seat. This process is called *solving for x*, and it forces the answer to appear, as if by magic, in the right seat. You just have to be careful to keep that seesaw balanced the whole time.

Adding and Subtracting

Take a look at the equation $x - 5 = 11$. It's basically asking "What number, if you subtract 5, gives you 11?" The answer is simple: 16. But let me show you the official mathematical way to get there.

Remember, your final goal is to get x all by itself on the left side of the equation, which means making that -5 next to it disappear. Well, it won't actually *disappear*, but you can turn it into 0 by adding it to its opposite: $-5 + 5 = 0$ (remember the additive inverse property from Chapter 1?). However, adding 5 to the left side of the equation will ruin the seesaw's balance—the (suddenly heavier) left side will come crashing to the ground, possibly causing a nasty-looking nosebleed.

To keep everything balanced, you'll have to add 5 to both sides of the equation/seesaw.

$$
\begin{array}{rcl}
x \quad - \quad 5 & = & 11 \\
\underline{+ \quad 5} & & \underline{+5} \\
x \quad + \quad 0 & = & 16 \\
x \qquad\qquad & = & 16
\end{array}
$$

Now that only x remains on the left side of the equation, the right side of the equation is the answer: 16. To make sure 16 is the right answer, go back to the original equation and replace x with 16:

$$x - 5 = 11$$
$$16 - 5 = 11$$
$$11 = 11$$

Since you get a true statement (11 definitely equals 11), your answer $x = 16$ is correct.

> **How'd They Do That?**
>
> By replacing x with 16 to check your answer, you're applying the substitution property (another property to add to your list), which allows you to substitute values for one another, as long as they are equal.

> **You've Got Problems**
>
> Problem 1: Solve the equation $8 + x = 19$ for x.

Multiplying and Dividing

In your endeavors to solve equations by isolating x's, you'll often have to eliminate numbers actually *attached* to the x, like in the equation $5x = 45$. The 5 and x are not

Talk the Talk _____

A number written next to a variable is called a **coefficient**. In the expression $2x$, for example, 2 is the coefficient. Note that coefficients are usually written in front of the variable: $2x$ instead of $x2$.

glued to one another or anything—they are just multiplied together and 5 is called the *coefficient* of x. Therefore the equation is actually asking "5 times what number is equal to 45?"

To answer that question, you can't add or subtract 5 like in the previous example, because neither addition nor subtraction can cancel out multiplication. You've got two options when it comes to eliminating coefficients:

- **Integer and decimal coefficients:** If the coefficient of x is not a fraction, divide both sides of the equation by that number.

- **Rational coefficients:** If x has a fractional coefficient $\dfrac{a}{b}$, multiply both sides of the equation by the reciprocal $\left(\dfrac{b}{a}\right)$ to solve for x. This may look familiar, because it's based on the multiplicative inverse property from Chapter 1.

Both of these techniques change the coefficient of x into some fraction $\dfrac{c}{c}$. In other words, x's coefficient will be a fraction with the same number up top and down below, and any fraction like that is equal to 1. (Any nonzero number divided by itself equals 1.) Therefore, the left side of the equation can be rewritten as $1x$, which means the exact same thing as plain old x, so your isolation procedure is complete.

Example 1: Solve the equations and verify the solutions.

(a) $5x = 45$

   Solution: The coefficient of x is the integer 5, so divide both sides of the equation by 5.

$$\frac{5x}{5} = \frac{45}{5}$$
$$x = 9$$

   To check the answer, substitute it for x in the original equation.

$$5x = 45$$
$$5(9) = 45$$
$$45 = 45$$

(b) $\frac{2}{3}y = 12$

Solution: The coefficient of x is a fraction, so multiply both sides of the equation by its reciprocal, $\frac{3}{2}$.

$$\frac{3}{2} \cdot \frac{2}{3}y = \frac{3}{2} \cdot \frac{12}{1}$$

$$\frac{6}{6}y = \frac{36}{2}$$

$$y = 18$$

To check the answer, substitute 18 for y in the original equation.

$$\frac{2}{3}y = 12$$

$$\frac{2}{3}\left(\frac{18}{1}\right) = 12$$

$$\frac{36}{3} = 12$$

$$12 = 12$$

You've Got Problems

Problem 2: Solve the equation $-\frac{4}{5}w = 16$ for w.

Equations with Multiple Steps

Most of the time, solving equations requires more than a single step. For example, think about the equation on the seesaw from the beginning of the chapter: $3x - 2 = 19$. Not only is there a –2 on the same side of the equal sign as the x, but there's also a 3 clinging to that x, like a dryer sheet stuck to a pant leg. In order to isolate x (and therefore solve the equation), you'll have to get rid of both of those numbers, using the techniques I've discussed thus far. (It wouldn't hurt to throw in some fabric softener with static cling controller as well.)

If a solution requires more than a single step, here's the order you should follow:

1. **Simplify the sides of the equation separately.** Each of the items added to or subtracted from one another in the equation are called terms. If two terms have the exact same variable portion, then they are called *like terms*, and you can combine them as though they were numbers.

Talk the Talk

Consider the expression $3y - 7y$. Since $3y$ and $-7y$ both have the exact same variable part (y), they are called **like terms,** and you can simplify by combining the coefficients while leaving the common variable alone: $3y - 7y = -4y$. (I'll discuss like terms in greater detail later on in Chapter 10.)

2. **Isolate the variable.** Using addition and subtraction, move all terms containing the variable you're isolating to one side of the equation (usually the left) and move everything else to the other side (usually the right). You're finished when you have something that looks like this: $ax = b$ (some number times the variable is equal to some other number).

3. **Eliminate the coefficient.** If the coefficient is something other than 1, divide by it (if it's an integer or a decimal) or multiply by its reciprocal (if it's a fraction).

Equation solving requires practice, and it's going to take some trial and error before you get good at it. Don't forget to check your answers!

Example 2: Solve each equation.

(a) $3x - 2 = 19$

Solution: You can't combine the terms on the left side of the equation, because $3x$ and -2 don't have the same variables (and therefore aren't like terms). That means your first objective is to isolate the variable term by adding 2 to both sides of the equation.

$$
\begin{array}{rcr}
3x \;-\; 2 & = & 19 \\
+\; 2 & & +2 \\
\hline
3x \;+\; 0 & = & 21
\end{array}
$$

Divide both sides by 3 to eliminate the coefficient.

$$\frac{3x}{3} = \frac{21}{3}$$
$$x = 7$$

(b) $-14 = 2x + 4(x + 1)$

Solution: You can do a bit of simplifying on the right side of the equation. Start by distributing 4 into the parentheses.

$$-14 = 2x + 4 \cdot x + 4 \cdot 1$$
$$-14 = 2x + 4x + 4$$

Combine the like terms $2x$ and $4x$.

$$-14 = 6x + 4$$

At this point, the problem looks a lot like the equation from part (a), except the variable term appears on the right side of the equation. There's no problem with that—it's perfectly fine. In fact, if you leave the $6x$ on the right side, it's easier to isolate the variable term. Just subtract 4 from both sides.

$$
\begin{array}{rcccc}
-14 & = & 6x & + & 4 \\
-4 & & & - & 4 \\
\hline
-18 & = & 6x & + & 0
\end{array}
$$

Divide both sides by 6 to eliminate the coefficient.

$$\frac{-18}{6} = \frac{6x}{6}$$
$$-3 = x$$

How'd They Do That?

In Example 2, part (b), I solved the equation by isolating x on the right side, rather than the left side. To tell you the truth, I prefer x on the left side as a matter of personal taste, even though it doesn't affect the answer at all.

According to the *symmetric property*, you can swap sides of an equation without affecting its solution. In other words, I could have flip-flopped the sides of the equation in 2(b) to get $2x + 4(x + 1) = -14$. If you solve that equation, you'll get $x = -3$, the exact same answer. So, go ahead and reverse the sides of an equation whenever you want.

(c) $-3(x + 7) = -2(x - 1) + 5$

Solution: Apply the distributive property to simplify both sides of the equation.

$$-3(x) + (-3)(7) = -2(x) + (-2)(-1) + 5$$
$$-3x - 21 = -2x + 2 + 5$$

Simplify the right side by combining the 2 and 5 (which are technically like terms, since they have the exact same variable part—no variables at all).

$$-3x - 21 = -2x + 7$$

Now, it's time to isolate the variable term. Do this by adding $2x$ to both sides (to remove all x-terms from the right side of the equation) and adding 21 to both sides as well (to remove non-x-terms from the left side of the equation).

$$
\begin{array}{rcrcr}
-3x & - & 21 & = & -2x & + & 7 \\
+2x & + & 21 & & +2x & + & 21 \\
\hline
-x & & & = & & & 28
\end{array}
$$

Critical Point

As demonstrated in Example 2, part c, a negative variable like $-w$ technically has a coefficient of -1, so you can rewrite it as $-1w$ if you want to. (This is similar to implied exponents and powers, where a plain old variable like w has an implied coefficient of 1, so $w = 1w^1$.)

At this point, you have $-x = 28$, which means "the opposite of the answer equals 28." Therefore, the correct answer is $x = -28$ (since -28 is the opposite of 28).

Here's another way to get the final answer. Rewrite the final line of the equation with a coefficient of -1 and then divide by that coefficient.

$$
-x = 28
$$

$$
\frac{-1x}{-1} = \frac{28}{-1}
$$

$$
x = -28
$$

(d) $y + 3 = \dfrac{1}{4} y + 5$

Solution: Neither side can be simplified because neither contains like terms, so skip right to isolating the variable terms. Subtract $\dfrac{1}{4} y$ and 3 from both sides of the equation. To subtract $y - \dfrac{1}{4} y$, remember that y and $1y$ mean the same thing. That lets you rewrite $y - \dfrac{1}{4} y$ as $1y - \dfrac{1}{4} y$. Subtract the like terms using a common denominator: $\dfrac{4}{4} y - \dfrac{1}{4} y = \dfrac{3}{4} y$.

$$
\begin{array}{rcrcr}
y & + & 3 & = & \dfrac{1}{4} y & + & 5 \\
-\dfrac{1}{4} y & - & 3 & & -\dfrac{1}{4} y & - & 3 \\
\hline
\dfrac{3}{4} y & & & = & & & 2
\end{array}
$$

The coefficient is fractional, so multiply both sides by its reciprocal.

$$\frac{4}{3}\left(\frac{3}{4}y\right)=\left(\frac{4}{3}\right)\frac{2}{1}$$

$$\frac{12}{12}y=\frac{8}{3}$$

$$y=\frac{8}{3}$$

Because 8 and 3 have no common factors (other than 1) the improper fraction cannot be simplified, so that's the final answer.

You've Got Problems

Problem 3: Solve the equations for x:
  (a) $3(2x - 1) = 14$
  (b) $2x - 7 = 4x + 13$

Absolute Value Equations

If you're trying to solve an equation that contains a variable trapped inside absolute value bars, you'll need to modify your technique slightly. That's because absolute value equations may have two answers instead of just one.

Here's what to do if you encounter an equation whose poor, defenseless variable is trapped in absolute value bars:

1. **Isolate the absolute value expression.** Before, you'd isolate the variable. This time, isolate the entire absolute value expression. Follow the same steps: start by adding or subtracting things out of the way and finish by eliminating a coefficient, if the expression has one.

2. **Create two new equations.** Here's the tricky part. You're actually going to make two new equations out of the original absolute value equation. The first equation should look just like the original, just without the bars on it. The second should look just like the first, except you'll take the opposite of the right side of the equation. This might sound tricky, but trust me, it's easy.

Critical Point

Even simple absolute value equations may have multiple solutions. Take the equation $|x| = 3$, for example. Both $x = 3$ and $x = -3$ make that equation true if you plug them in.

3. **Solve the new equations to get your answer(s).** Both of the solutions you get are answers to the original absolute value equation.

To remind myself that absolute value equations require two separate parts, I sometimes imagine that those absolute value bars are little bars of dynamite that blow the original equation into two pieces. Now that you know the process involved, let me show you how to handle the explosives correctly.

Example 3: Solve the equation $4|2x-3|+1=21$.

Solution: Start by subtracting 1 from both sides of the equation in order to isolate the absolute value quantity on the left.

$$4|2x-3| = 20$$

To complete the isolation process, divide both sides by 4.

$$\frac{4|2x-3|}{4} = \frac{20}{4}$$

$$|2x-3| = 5$$

Now that only the absolute values remain on the left side, it's time to create two new equations. The first looks just like the above equation (without bars attached): $2x - 3 = 5$. The second equation looks almost the same, except the number on the right side will be the opposite: $2x - 3 = -5$. Solve the equations separately.

$$
\begin{array}{ll}
2x - 3 = 5 & \qquad 2x - 3 = -5 \\
2x = 8 & \qquad 2x = -2 \\
x = 4 & \qquad x = -1
\end{array}
$$

The answer is "$x = -1$ or $x = 4$." If you find two answers hard to swallow, watch what happens when you plug them into the original equation.

$$
\begin{array}{ll}
4|2x-3|+1=21 & \qquad 4|2x-3|+1=21 \\
4|2(-1)-3|+1=21 & \qquad 4|2(4)-3|+1=21 \\
4|-2-3|+1=21 & \qquad 4|8-3|+1=21 \\
4|-5|+1=21 & \qquad 4|5|+1=21 \\
4(5)+1=21 & \qquad 4(5)+1=21 \\
21=21 & \qquad 21=21
\end{array}
$$

You've Got Problems

Problem 4: Solve the equation $|x-5|-6=4$ for x.

Equations with Multiple Variables

So far, when I've asked you to solve an equation for a variable, it was pretty obvious which one I was talking about. For example, to solve the equation $3x + 2 = 23$, you'd solve for (isolate) the x variable. Why? Because it's the only variable in there! That x is enjoying all the attention, like the only girl in an all-boys school.

I need to add another skill to your equation-solving repertoire that will be extremely important in Chapter 5, solving for a variable when there's more than one variable in the equation. Don't worry—it's basically done the same way you've solved the other equations in this chapter. Here's the only difference: when you're finished, you'll have variables on both sides of the equation, not just one side.

Example 4: Solve the equation $-2(x - 1) + 4y = 5$ for y.

Solution: Start by simplifying the left side of the equation.

$$-2x + 2 + 4y = 5$$

Now it's time to separate the variable term. Since you're trying to isolate y, eliminate every term not containing y from the left side of the equation. In other words, add $2x$ to, and subtract 2 from, both sides of the equation.

$$\begin{array}{rcccccc} -2x & + & 2 & + & 4y & = & 5 \\ +2x & - & 2 & & & & +2x & - & 2 \\ \hline & & & & 4y & = & 2x & + & 3 \end{array}$$

It might have felt weird to move the $-2x$ term to the right side of the equation, but it had to go—you're solving for y, not x. All that's left to do now is eliminate the y-coefficient, so divide both sides by 4.

$$\frac{4y}{4} = \frac{2x+3}{4}$$

$$y = \frac{2x+3}{4}$$

Even though that answer is correct, you need to know another way to write it. (This will come in handy in Chapter 5.) Any time two or more things are added or subtracted in the numerator of a fraction, you can break that fraction into smaller fractions. Each term of the numerator, in this case $2x$ and 3, gets its own fraction with a copy of the denominator.

$$y = \frac{2x}{4} + \frac{3}{4}$$

$$y = \frac{1x}{2} + \frac{3}{4}$$

Variables in the numerator of a fraction can be written beside the fraction, so move that x.

$$y = \frac{1}{2}x + \frac{3}{4}$$

You've Got Problems
Problem 5: Solve the equation $9x + 3y = 5$ for y.

The Least You Need to Know

◆ In order to solve equations, you must keep them balanced at all times—perform the operations on both sides of the equation simultaneously.

◆ To solve an equation for a variable, isolate the variable on one side of the equal sign.

◆ Check your solutions by plugging them back into the original equations.

◆ In order to solve equations with variables inside absolute value signs, you have to create two new equations that probably have different answers.

5

Graphing Linear Equations

In This Chapter

- ♦ Navigating the coordinate plane
- ♦ Plotting points and graphing lines
- ♦ Identifying the slope of a line
- ♦ Graphing linear absolute values

I am a horrible artist. Although I am pleased with my understanding of math, I would give my left arm for the ability to draw something as meaningless as a realistic-looking chicken. (My feeble attempts at a chicken always end up looking something like a polar bear balanced on top of a four-slot toaster.)

Luckily, mathematical drawings are much more technical (read: boring) than creative drawings, so I can handle them. In this chapter, I will focus on drawing mathematical graphs—pictures of things like equations, but not things like arctic animals on food service appliances. By the end of this chapter, you'll be able to draw pictures of linear equations, whose graphs are plain old straight lines. Nothing's easier to draw than a straight line, right? (Even though my attempts to draw straight lines end up looking like chickens.)

Climb Aboard the Coordinate Plane

Thanks to a fun-loving mathematician named René Descartes (a guy who contributed so much to math that I won't tease him for having a girl's name), we have the vastly useful mathematical tool called the *coordinate plane*. It is basically a big, flat grid used to visualize mathematical graphs. Check out Figure 5.1 for a look at the coordinate plane.

Figure 5.1

The horizontal x-axis and the vertical y-axis intersect at the origin, and divide the plane into four quadrants (which are labeled with Roman numerals).

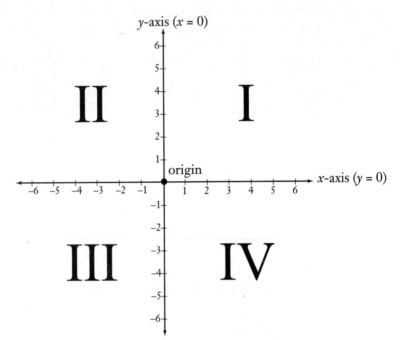

To understand the coordinate plane, pretend that it is the map of a small town. In this town, there are only two main roads, one that runs horizontally (called the *x-axis*) and one that runs vertically (called the *y-axis*). These two roads intersect once, right in the middle of town, at a location called the *origin*. This intersection splits the town into four *quadrants*, which are numbered in a very specific way. The northeast section of town is quadrant one (I), the northwest is quadrant two (II), the southwest is quadrant three (III), and the southeast is quadrant four (IV).

Talk the Talk

The **coordinate plane** is a flat grid used to visualize mathematical graphs. It is formed by the **x-axis** and **y-axis**, a horizontal and a vertical line that meet at a point called the **origin**. The axes split the plane into four **quadrants**.

Just like the road "Main Street" might go by a more formal name like "State Route 4" in the real world, the x- and y-axes have official addresses in addition to their names. The address of the x-axis is the equation $y = 0$. In fact, every horizontal road in this town has the address "$y = some\ number$." The first horizontal road above the x-axis, for example, has equation (address) $y = 1$; the road above that has equation $y = 2$; and so on. The first horizontal road *below* the x-axis has equation $y = -1$, the next is $y = -2$, etc. Basically, all of the positive horizontal street addresses occur in quadrants I and II, and the negative streets are located in quadrants III and IV.

In a similar way, the vertical streets have equations that look like "$x = some\ number$." The y-axis has equation $x = 0$, and the streets to its right are numbered $x = 1$, $x = 2$, $x = 3$, etc. The streets left of the y-axis have equations $x = -1$, $x = -2$, $x = -3$, etc. So, positive vertical streets run through quadrants I and IV, and the negative vertical streets run through quadrants II and III.

Because all of the streets have these handy addresses, it's very simple to pinpoint any location in town. For example, let's say I find a great bakery at the intersection of streets $x = -3$ and $y = 4$, as shown in Figure 5.2.

Critical Point

You can scale each grid mark on the coordinate plane—each line could represent 1, 2, 5, 10, or any fixed number of units. However, in this book, each mark usually represents a distance of 1.

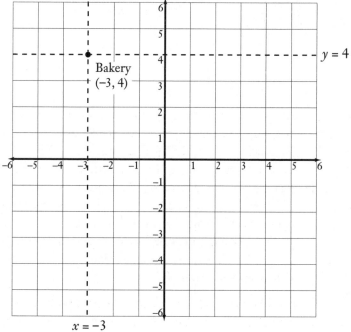

Figure 5.2

You're going to love the fresh-baked bagels in this quadrant II bakery.

Talk the Talk

Every point on the coordinate plane is described by a **coordinate pair** (x,y). The x portion of the coordinate pair is sometimes called the abscissa, and the y portion is called the ordinate, but that terminology is very old, antiquated, and formal, so you may not hear it unless your teacher is old and antiquated.

Written on the door of the bakery is its official coordinate plane address: (–3,4). You see, every location in the coordinate plane has an address (x,y) called a *coordinate pair,* based on the intersecting street numbers. When writing the coordinate pair, make sure to list the x street first, followed by the y street. Once you get good at plotting (graphing) points on the coordinate plane, you'll be ready to do more advanced things, like connecting those dots to form graphs.

Example 1: Plot these points on the coordinate plane: $A = (2,0)$, $B = (0,-4)$, $C = (-3,-2)$, $D = \left(-\frac{7}{2}, 3\right)$, $E = (5,-1)$, and $F = (6,2)$.

Solution: Remember that each coordinate pair represents the intersection of a vertical street (the first number in the pair) and a horizontal street (the second number in the pair). For instance, point C lies at the intersection of the third vertical street to the *left* of the origin (because $x = -3$ is negative) and the second horizontal street *below* it (because $y = -2$ is negative).

Figure 5.3

The solution to Example 1. Grid lines are included on the coordinate plane so that the points are easier to identify.

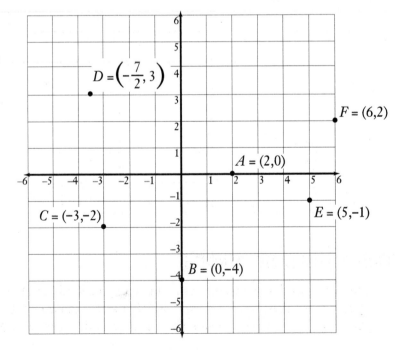

Points A and B fall on the x- or the y-axis, because they each contain a 0 in the coordinate pair. The trickiest point to plot is D, because it contains a fraction. To make things easier for you, convert the improper fraction $-\frac{7}{2}$ into the mixed number $-3\frac{1}{2}$ (using the technique from Chapter 2). To plot D, count three and a half units to the left of the origin and then three units up. All of the points are plotted in Figure 5.3.

You've Got Problems

Problem 1: Identify the points plotted on the coordinate plane in Figure 5.4.

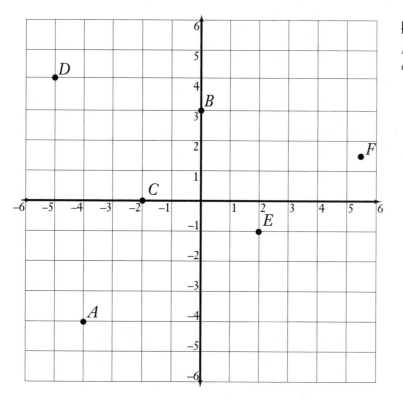

Figure 5.4

List the (x,y) coordinates for each of the points A through F.

Sketching Line Graphs

Not that plotting points isn't fun, but it gets old kind of fast. So let's up the ante and graph some lines, which are only slightly more complicated. Lines are the graphs of *linear equations*, which are equations that are in this form: $ax + by = c$. In other words, they're equations that contain x and y (usually with coefficients attached) and a number without a variable, called a *constant*.

Talk the Talk

A **linear equation** looks like $ax + by = c$, where x and y are variables, a and b are their coefficients, and c is a number with no variable attached (called a **constant**). Here are some examples of linear equations: $x - y = 5$, $3x - 2y = 1$, and $4y - 2 = x$. (The terms don't always have to be in the same order.)

How'd They Do That?

Is $x = 7$ a linear equation? Yes! Even though a linear equation looks like $ax + by = c$ and $x = 7$ doesn't have any y's in there, it's still linear because, technically, $b = 0$. In other words, there's no use writing the y-term because its coefficient is 0.

Back in Chapter 4, you solved linear equations like $2x - 1 = 15$; they're simpler because they only have one variable in them. It's pretty easy to figure out that the solution to the equation $2x - 1 = 15$ is $x = 8$, but when there are two variables in an equation, a nutty thing happens. Instead of just *one* solution, there are an *infinite* number of solutions!

Take a look at the linear equation $x + y = 9$, which translates to "What two numbers add up to 9?" Well, there are many pairs of numbers x and y that could make that statement true. If $x = 1$ and $y = 8$, you get $1 + 8 = 9$. How about $x = 11$ and $y = -2$? Those two numbers add up to 9 as well! How can you possibly give a solution to this equation when there are so many dang answers?

The best thing to do is to write both of those solutions as ordered pairs and plot them on a graph: $(1,8)$ and $(11,-2)$.

Here's the cool thing: the infinite number of solution points, not just the two points I pointed out—*all* of them, will make a perfectly straight line in the coordinate plane. Finally, math terminology that makes sense! They're called *linear* equations because their graphs are *lines*.

If you ever wondered why you had to graph equations, now you know! The graph is simply a visual representation of all the ordered pairs (x,y) that, if plugged back into the linear equation, would make it true. Now that you know *why* you have to make graphs, it's time to figure out *how* to make them.

Graphing with Tables

The easiest way to graph any equation is to use a table. Basically, you plug a whole bunch of numbers in for x and find out what the corresponding y-values are that complete the ordered pair. Once you do, you can plot the points on the coordinate plane, secure in the knowledge that they fall on the graph.

Here are the steps to follow to graph a linear equation using a table:

1. **Solve the equation for *y*.** This will streamline the process—since I just showed you how to do this at the end of the last chapter, why not show off your skills?

2. **Plug in a few values for *x* and record the resulting *y*-values.** Write each corresponding *x* and *y* pair together as a coordinate pair (x,y).

3. **Plot the points on the coordinate plane and connect them to form the graph.** In these early stages of graphing, I suggest you use graph paper. Later on, as you get more experienced, you'll be able to graph without it.

 **Kelley's Cautions**

How many *x*-values should you plug in? Well, geometry tells us that it takes two points to define a line, but if you plot three, you can use the third one to check yourself. If they don't all fall on the same line, then you made an algebra mistake somewhere.

Example 2: Sketch the graph of $2x - y = 5$ using a table.

Solution: Solve for *y* by subtracting $2x$ from both sides of the equation and then either multiplying or dividing everything by -1 to make *y* positive.

$$-y = -2x + 5$$
$$-1(-y) = -1(-2x + 5)$$
$$y = 2x - 5$$

Now it's time to construct the table. The left column will contain the *x*-values you'll plug in. (I usually choose -1, 0, and 1 since they are small, simple numbers.) The middle column is used to perform the calculations, and the right column is used to record the resulting coordinate pair.

x	$y = 2x - 5$	(x,y)
-1	$y = 2(-1) - 5$ $y = -2 - 5 = -7$	$(-1,-7)$
0	$y = 2(0) - 5$ $y = -5$	$(0,-5)$
1	$y = 2(1) - 5$ $y = 2 - 5 = -3$	$(1,-3)$

Now that you know the points $(-1,-7)$, $(0,-5)$, and $(1,-3)$ represent solutions to the equation plot them and connect the dots to get the graph (pictured in Figure 5.5).

Figure 5.5

The line $2x - y = 5$ will extend infinitely in each direction. Note that the x- and y-axes do not appear in the exact middle of this graph. Since all three of the points occur below the x-axis, the focus of the graph is shifted.

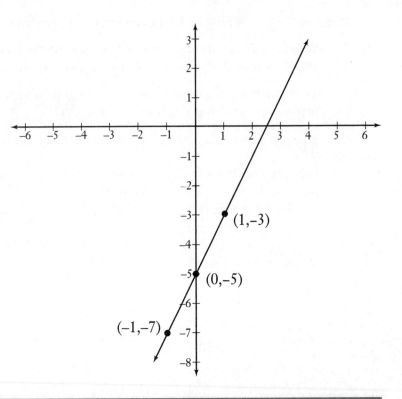

(1,–3)

(0,–5)

(–1,–7)

You've Got Problems

Problem 2: Sketch the graph of $-4x + y = 2$ using a table.

Graphing with Intercepts

The point (or points) at which a graph intersects the *x*- or *y*-axis is called an *intercept*. You may have noticed that coordinates on the axes always contain a 0. In fact, a coordinate on the *x*-axis will always have a *y*-value of 0, and a coordinate on the *y*-axis will always have an *x*-value of 0. This makes calculating intercepts easy. All you have to do to find an *x*-intercept is plug in 0 for *y*, and to calculate a *y*-intercept, plug in 0 for *x*. Once you've found those two intercepts, plot them and connect the dots to graph the line!

Talk the Talk

An **intercept** is a point on either the *x*- or *y*-axis that the graph passes through. Every *x*-intercept has the form (*x*,0) and every *y*-intercept has the form (0,*y*).

Example 3: Graph $3x - y = 6$ by calculating its intercepts.

Solution: If you plug in 0 for x and solve, you'll get the y-intercept.

$$3(0) - y = 6$$
$$-y = 6$$
$$y = -6$$

The y-intercept is $(0,-6)$. Now plug 0 in for y (back in the original equation) and solve to get the x-intercept.

$$3x - (0) = 6$$
$$3x = 6$$
$$x = 2$$

The x-intercept is $(2,0)$. Plot the intercepts and connect the points to get the graph, pictured in Figure 5.6.

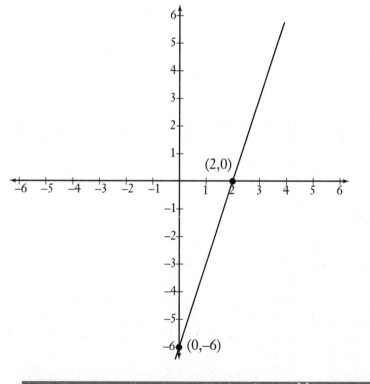

Figure 5.6

The solution to Example 3, the graph of $3x - y = 6$.

You've Got Problems

Problem 3: Graph $4x + 2y = -8$ by calculating its intercepts.

It's a Slippery Slope

The *slope* of a line is a number that describes how "slanty" that line is. Is the line steep or shallow? Does it rise from left to right or fall? Even if you don't have the graph of the equation, you can answer all of these questions knowing only the slope of the line.

Calculating Slope of a Line

As useful as the slope of a line is, it's remarkably easy to calculate. All you have to do is pick out two points on the line, which I'll call points A and B, for reference. The slope of the line is equal to this fraction:

$$\frac{\text{number of vertical units between } A \text{ and } B}{\text{number of horizontal units between } A \text{ and } B}$$

Let me explain what I mean in terms of a real-life example—placekicking in football. Whenever a kicker comes onto the field to try for an extra point or a field goal, he must make sure that his path to the ball is precise. He needs to get a running start at the ball to give his kick some momentum. (If he just stood there and swung his leg, the ball wouldn't go very far.) However, if his approach path is off course, the ball won't go exactly where he's intending it.

To make sure their approach paths are exactly right, most right-footed kickers do the same thing: they stand over the ball, take two steps backward, and then two steps to their left to reach their starting position. This creates an imaginary line of approach, which passes through two points: the position of the ball and their starting position, as shown in Figure 5.7.

Figure 5.7

To position himself correctly, the kicker takes two steps back from the ball and then two steps left.

Kicker's path

Imagine we're looking down at the field, perhaps from an airplane or a blimp advertising tires, and you can easily calculate the slope of that approach line. The slope's numerator is equal to the number of steps he takes forward or backward (up or down as we look down on the field); because he backs up two steps from the ball, the numerator will be –2. (If he'd have gone forward, and moved up according to our perspective, the number would have been positive.)

The denominator of the slope equals the number of steps he takes left or right, where left steps are negative and right steps are positive. Since he takes two steps left, the denominator equals –2. Therefore, the slope of the line is:

$$\frac{-2}{-2} = 1$$

Even though you can calculate the slope of a line by picking two of its points and counting the number of steps (or units) it takes to get from one to the other, it's not a very handy technique, especially if the points' coordinates aren't integers.

There is a handy formula you can use to calculate the slope of a line, given the coordinates of two of its points—the formula basically counts the horizontal and vertical changes for you. It works like this: If a line contains points (a,b) and (c,d), then the slope of the line is equal to:

$$\frac{d - b}{c - a}$$

Critical Point _____

You can also calculate the slope of the kicker's approach by going from the kicker to the ball, rather than vice versa. In that case, the numerator is 2 (two steps forward or up) and the denominator is 2 (two steps right). That makes the slope $\frac{2}{2} = 1$. Either way, you get the same answer.

Kelley's Cautions _____

Make sure to subtract the y-values in the numerator and the x-values in the denominator. Students often reverse them, mistakenly placing the x's in the top of the fraction, so be cautious.

Technically speaking, slope is defined as the change in y (the difference of two y-values) divided by the change in x (the difference of two x-values).

Example 4: Calculate the slope of the line that passes through the points (4,–1) and (–3,9) and explain what it means.

Solution: Apply the slope formula—subtract the y-values of the points and divide that by the difference of the x-values.

$$\frac{9-(-1)}{-3-(4)} = \frac{9+1}{-3-4} = \frac{10}{-7} = -\frac{10}{7}$$

Remember, if a fraction is negative, you can rewrite the fraction, placing the negative sign in the denominator or the numerator, so $-\frac{10}{7} = \frac{-10}{7} = \frac{10}{-7}$.

What do these slopes mean? A slope of $\frac{-10}{7}$ means that you can go down 10 units and right 7 units from one point on the line to reach another point on that line, because that's exactly what you do to travel from (–3,9) to (4,–1).

Alternatively, the slope $\frac{10}{-7}$ assures you that you can go up 10 units and left 7 units from one point on the line to reach another point, because that's the path from (4,–1) to (–3,9).

You've Got Problems

Problem 4: Calculate the slope of the line passing through the points (4,0) and (–5,6).

Interpreting Slope Values

You can tell a lot about a line just by looking at its slope. Here are a few facts about the gripping, romantic relationship between a line and its slope:

♦ **Lines with positive slopes rise.** If a line's slope is greater than 0, its graph will slant upwards from left to right.

♦ **Lines with negative slopes fall.** If its slope is less than 0, a line slants downward from left to right.

♦ **Lines with large slopes are steep.** The further a slope's absolute value is from 0, the steeper its graph will be. A line with slope 2 is significantly steeper than a line with slope 1, so the steepness increases quickly as slope values increase. A line with a slope of –5 is also very steep—it's just slanted in a different direction.

♦ **Lines with small slopes rise more gradually.** The closer a slope's absolute value is to 0, the more its graph will resemble a horizontal line.

♦ **Horizontal lines have slopes equal to 0.** Any two points on a horizontal line have the same *y*-value, so when you calculate the slope of the line passing through those points, the numerator equals 0.

♦ **Vertical lines have undefined slopes.** Learn more about this in the nearby sidebar.

Critical Point _____

What does it mean for a vertical line to have an *undefined* slope? Basically, the slope of a vertical line is an illegal number. Take, for example, the vertical line $x = 3$. Select any two points from the line, like $(3,4)$ and $(3,9)$, and plug them into the slope formula.

$$\frac{9-4}{3-3} = \frac{5}{0}$$

Did you know that you're not allowed to divide by 0 in math? It's as illegal as ripping those little tags off your mattress.

Any fraction with 0 in the denominator (but not in the numerator) is classified as *undefined*. That's why your textbook says the slope of a vertical line is undefined, or that the line has "no slope."

How can you remember that vertical slopes are undefined but horizontal slopes are 0? Think about trying to walk along the lines. How much effort would it take to walk along a horizontal line? Zero.

How much effort would it take to walk along a vertical line? You can't! It's impossible, because there is no slope in a vertical line. It'd be like trying to walk up a wall, and unless you have been bitten by a mutant spider, wear brightly colored red-and-blue spandex, and have a girlfriend named Mary Jane, that skill's probably not on your resumé.

Kinky Absolute Value Graphs

One final word about graphing linear equations before I wrap this chapter up. Remember in Chapter 4 when you learned that equations containing an *x* in absolute values required a slightly different solution technique than non-absolute value equations? You had to break the equation into two parts to get the answer. Well, you have to graph them a little differently than regular linear equations, too. Whereas the graph of a normal linear equation looks like a line, an absolute value linear equation graph looks like a "V." It's basically a line with a kink in it, a sharp point (or *vertex*) where the graph changes directions.

Look at Figure 5.8, which contains the graphs of $y = x - 3$ and $y = |x - 3|$. Both of the graphs look the same to the right of the line $x = 3$. However, when x is less than 3, the linear graph dips below the x-axis (meaning that its y-values are negative). See how the right-hand graph makes a sharp turn in order to avoid going below the x-axis? If $y = |x - 3|$, then y (like any other absolute value) cannot be negative, so the graph avoids negative y-values like the plague.

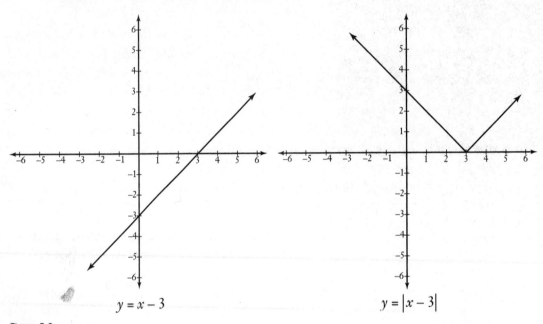

$$y = x - 3 \qquad\qquad y = |x - 3|$$

Figure 5.8

The graph of $y = |x - 3|$ takes drastic measures to avoid negative values, unlike the graph of $y = x - 3$.

The best way to graph a linear absolute value is to figure out what the vertex is, plot it, and then plot one point to the right and one point to the left of it (to get the branches of the graph).

Example 5: Graph the equation $y = -\dfrac{1}{2}|x + 4| - 3$.

Solution: To find the vertex of the graph, set the contents of the absolute values equal to 0 and solve for x.

$$x + 4 = 0$$
$$x = -4$$

Now plug $x = -4$ back into the original equation and solve for y.

$$y = -\frac{1}{2}|(-4)+4|-3$$

$$y = -\frac{1}{2}(0)-3$$

$$y = -3$$

Therefore, the vertex of the graph is the point (–4,–3), so plot that point on the coordinate plane. Now choose one *x*-value to the left of the vertex and one to the right (in other words, choose one *x* which is less than –4 and one which is greater), and plug them both into the original equation. I'll chose *x* = –6 and *x* = –2.

$$y = -\frac{1}{2}|(-6)+4|-3 \qquad y = -\frac{1}{2}|(-2)+4|-3$$

$$y = -\frac{1}{2}|-2|-3 \qquad y = -\frac{1}{2}|2|-3$$

$$y = -\frac{1}{2}(2)-3 \qquad y = -\frac{1}{2}(2)-3$$

$$y = -1-3 \qquad y = -1-3$$

$$y = -4 \qquad y = -4$$

Plot the resulting coordinate pairs, (–6,–4) and (–2,–4), each time drawing a line that begins at the vertex and passes through one of the points, as illustrated by Figure 5.9.

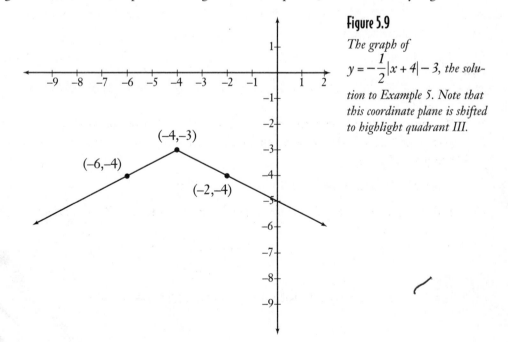

Figure 5.9

The graph of
$$y = -\frac{1}{2}|x+4|-3, \text{ the solu-}$$
tion to Example 5. Note that this coordinate plane is shifted to highlight quadrant III.

Kelley's Cautions _____

If you draw an absolute value graph and it dips below the x-axis, that doesn't necessarily mean you graphed it wrong! Some linear absolute value equations will *still* have negative y's. For example, if you plug $x = 2$ into the equation $y = |x| - 5$, you'll get $y = -3$. I'll discuss this in greater detail in Chapter 16.

You've Got Problems

Problem 5: Graph the equation $y = |2x - 4| + 1$.

The Least You Need to Know

♦ The coordinate plane is a grid formed by the intersection of a horizontal line called the x-axis and a vertical line called the y-axis.

♦ Vertical lines have equation "$x = c$," and horizontal lines have equation "$y = d$," where c and d are real numbers.

♦ You can graph linear equations by plotting points or calculating their x- and y-intercepts.

♦ The slope of a line is a ratio that describes how "slanty" a line is by calculating the vertical and horizontal change between points on the line.

♦ The slope of a horizontal line is 0, and a vertical line has an undefined slope.

♦ The graph of a linear absolute value equation has a sharp point in it called the "vertex."

Chapter 6

Cooking Up Linear Equations

In This Chapter

♦ Determining equations of lines

♦ Applying the point-slope and slope-intercept formulas

♦ Writing linear equations in standard form

♦ Exploring connections to geometry

Now that you can graph lines and have some idea of what a slope is, it's time to get down and dirty with linear equations and create them from scratch. Even if you're not very good at cooking or baking, don't worry. The recipes for linear equations are pretty simple, require very few ingredients, and almost never result in an oven fire.

All you need to create a linear equation is its slope and one of the points on the line. Once you've got those two pieces of information, you'll be cranking out equations as fast as Sarah Lee can whip out pound cakes. Along the way, you'll even get a better understanding of linear equations and how all of their pieces fit together correctly.

Point-Slope Form

If you know the slope of a line and one of the points on that line, you can use the *point-slope formula* to figure out what the equation of that line is.

Point-slope formula: If a line has slope m and passes through the point (x_1, y_1), then the equation of the line is:

$$y - y_1 = m(x - x_1)$$

Are you wondering where that m came from? For some reason, math people have used the variable m to represent slope for a long time. Believe it or not, no one quite knows why. I could wax historical about this mathematical mystery, but it's really not all that interesting. All you need to know is that m is the variable used to represent slope in every linear equation formula you'll see in this book.

Critical Point

Remember, no matter which of the techniques described in this chapter you use to find a linear equation, you always need two things: the slope of the line and a point on the line.

Critical Point

Subscripts, like the little 1 on x_1 and y_1 in the point-slope formula, are there to help you figure out which x's and y's you're talking about (because the formula has two x's and two y's in it).

To create a linear equation, plug a slope in for m, the x-value from one of the ordered pairs for x_1, and the matching y-value for y_1; then simplify.

Example 1: Write the equation of the line with slope –3 that passes through the point (–1,5) and solve the equation for y.

Solution: Because the slope of the line is –3, set $m = -3$ in the point-slope formula. You should also replace x_1 with the known x-value (–1) and replace y_1 with the matching y-value (5).

$$y - y_1 = m(x - x_1)$$
$$y - (5) = -3(x - (-1))$$

Simplify the right side of the equation.

$$y - 5 = -3(x + 1)$$
$$y - 5 = -3x - 3$$

The problem asks you to solve for y, so isolate y on the left side of the equation by adding 5 to both sides.

$$y = -3x + 2$$

That's all there is to it! There are two quick things you can do to check your answer. First, make sure that the slope of the line is its *x*-coefficient. (More on this in the next section.) Second, plugging $x = -1$ and $y = 5$ into the equation must produce a true statement.

$$y = -3x + 2$$
$$5 = -3(-1) + 2$$
$$5 = 3 + 2$$
$$5 = 5$$

You've Got Problems

Problem 1: Write the equation of the line with slope 4 that passes through the point (2,–7) and solve the equation for *y*.

Slope-Intercept Form

Are you wondering why you had to solve the linear equations in Example 1 and Problem 1 for *y*? I wasn't just making you jump through hoops (although I am impressed by your agility, I must admit); there's actually a very good reason to do it. Once a linear equation is solved for *y*, it is in *slope-intercept form*.

There are two big reasons to put an equation in slope-intercept form (and you can probably figure out what they are based on the name): you can identify the *slope* and the *y-intercept* of the line (get this) without doing any additional work at all! Plus, writing equations in slope-intercept form while you do sit-ups could actually give you great-looking abs!

 Talk the Talk

Once a linear equation is solved for *y*, it is in **slope-intercept form**: $y = mx + b$. The coefficient of the *x*-term (*m*) is the slope of the line and the number (or **constant**) *b* is the *y*-intercept.

Let's get mathematical for a moment. Technically, the slope-intercept form of a line is written like this:

$y = mx + b$, where *m* is the slope and *b* is the *y*-intercept

In other words, once you solve a linear equation for *y*, the coefficient of *x* is the slope of the line, and the number with no variable attached (called the *constant*) marks the spot on the *y*-axis, (0,*b*), where the line passes through.

Example 2: Identify the slope and the coordinates of the *y*-intercept for the linear equation $x - 4y = 12$.

Solution: All you have to do to transform an equation into slope-intercept form is to solve it for *y*. To isolate the *y*, subtract *x* from both sides of the equation and then divide everything by its coefficient –4.

$$-4y = -x + 12$$

$$y = \frac{1}{4}x - 3$$

The *x*-term's coefficient is $\frac{1}{4}$, so the slope of the line is $\frac{1}{4}$. The constant is –3, so the graph of the line will pass through the *y*-axis at the point (0,–3). (Don't forget that the *x*-coordinate of a point on the *y*-axis is always 0.)

You've Got Problems

Problem 2: Identify the slope and the coordinates of the *y*-intercept for the linear equation 3*x* + 2*y* = 4.

Graphing with Slope-Intercept Form

Chapter 5 already demonstrated a couple of ways to graph linear equations, but I thought I'd toss one more method into the mix. I know you don't need a hundred different ways to graph lines any more than you need a hundred different ways to tie your shoes, but because graphing using the slope-intercept form of a line is my favorite technique, I want to share it with you. It'll be a bonding moment for us. Besides, there's nothing more boring than plotting point after point to come up with a graph. This way is a little different, and it helps you understand how the slope works, in case you are a little fuzzy on that.

Here's how to graph a line using slope-intercept form:

1. **Solve the equation for *y*.** The equation needs to be in slope-intercept form.

2. **Determine the slope and *y*-intercept of the line.** Remember, this is as easy as looking at the numbers on the right side of the slope-intercept equation.

3. **If the slope is negative, place the negative sign in either the numerator or the denominator.** You can stick the negative sign in either place (but not both at the same time), it doesn't matter which.

4. **Plot the *y*-intercept of the graph.** Remember, the *y*-intercept has coordinates (0,*b*); plot that point.

5. **Starting at the y-intercept, use the slope to count steps to another point on the graph.** Positive numbers in the numerator mean up, and negative numbers in the numerator mean down; positives in the denominator mean right, and negatives in the denominator mean left. For example, a slope of $\frac{-2}{3}$ means to count down 2 units and right 3 units from the y-intercept to get another point on the graph.

6. **Connect the two points.** Just connect the dots to finish the graph. Remember, it extends infinitely in both directions.

How'd They Do That?

Are you wondering where the slope-intercept form comes from, and how we can be so sure that the x-coefficient is the slope and b is the y-intercept?

Let's say there's a line with slope m and y-intercept $(0,b)$ (the familiar variables from slope-intercept form). Use the point-slope formula to get the equation of that line.

$$y - y_1 = m\,(x - x_1)$$
$$y - b = m\,(x - 0)$$
$$y - b = mx$$

Solve for y and you end up with the slope-intercept form of a line.

$$y = mx + b$$

A star is born!

Example 3: Graph the equation $5x + 3y = 12$ using slope-intercept form.

Solution: Start by solving the equation for y.

$$3y = -5x + 12$$
$$y = -\frac{5}{3}x + 4$$

The y-intercept of the line is $(0,4)$, and its slope should be rewritten with the negative sign in either the numerator or denominator: $\frac{-5}{3}$ or $\frac{5}{-3}$.

Plot the y-intercept and count your way to the next point based on the slope you chose. You should either count down five and right three units $\left(\frac{-5}{3}\right)$ or count up five and left three units $\left(\frac{5}{-3}\right)$. Either way, when you connect the points, you'll end up with the same line, illustrated in Figure 6.1.

I like this method of graphing lines because it sort of feels like you're reading a treasure map: start at the big palm tree, and then take five paces due south and three paces due east to reach the Golden Booty (which, by the way, sounds like a great name for an R&B star).

Figure 6.1

You can either move up 5 steps and left 3 steps from the y-intercept (0,4), or down 5 steps and right 3 steps from the y-intercept to reach another point on the line.

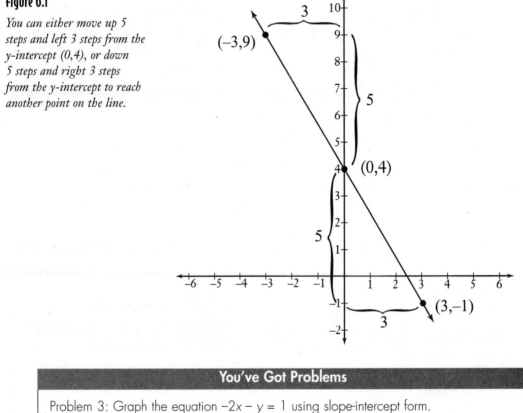

You've Got Problems

Problem 3: Graph the equation $-2x - y = 1$ using slope-intercept form.

Standard Form of a Line

As long as you keep an equation balanced, you can do just about anything to both of its sides to change the way it looks. For example, take a look at these two linear equations:

$$3x - 2y = 4 \qquad 1 + \frac{1}{2}y = \frac{3}{4}x$$

In a few steps, you can take that ugly, fraction-riddled equation on the right and transform it into the much more attractive (and confident) equation on the left. Start by multiplying both sides of the equation by 4.

$$\frac{4}{1}\left(1+\frac{1}{2}y\right)=\frac{4}{1}\left(\frac{3}{4}x\right)$$

$$4+\frac{4}{2}y=\frac{12}{4}x$$

$$4+2y=3x$$

Now flip-flop the sides of the equation, which is okay according to the symmetric property.

$$3x = 4 + 2y$$

Subtract $2y$ from both sides and the transformation is complete.

$$3x - 2y = 4$$

Isn't that much better-looking than $1+\frac{1}{2}y=\frac{3}{4}x$? Most math teachers think so. In fact, they feel so strongly about it that they usually require you to write your answers in that tamer, less-fractiony format.

This prettier version of the equation is called *standard form*, and it has the following properties:

◆ The equation looks like $ax + by = c$; in other words, the x- and y-terms are on the left side of the equation and the constant is on the right side.

◆ All of the numbers (a, b, and c) are integers.

◆ The coefficient of the x-term, a, is positive.

Why bother putting equations in standard form? To be honest, I'm not sure why teachers stress this so much. I much prefer slope-intercept form, because its values m and b

Talk the Talk

An equation in **standard form** looks like $ax + by = c$, where b and c are integers and a is a positive integer.

Critical Point

If a linear equation is in standard form $ax + by = c$, you can use the shortcut formula $m = -\frac{a}{b}$ to calculate the slope. For example, the slope of $5x - 8y = -2$ is

$m = -\frac{a}{b} = -\frac{5}{-8} = \frac{5}{8}$.

actually represent something, whereas the coefficients of standard form aren't all that useful.

Some say that standard form is important because every single linear equation can be put into standard form, which is not true of slope-intercept form. (A vertical line, like $x = 2$, can't be solved for y, and if there's no y around, that means slope-intercept form is not an option.) Maybe that's true, but I think standard form is generally preferred because people hate fractions—even your algebra teacher, although she'd never admit it.

Example 4: Put the linear equation $-\frac{2}{3} - 4x = \frac{5}{9}y$ in standard form.

Solution: The first order of business is to get rid of all those ugly fractions. Take a look at the denominators in the equation (3 and 9) and calculate the least common denominator. (In case you forgot how to do this, review the process in Chapter 2—it had something to do with a number named Bubba.)

In this case, the least common denominator is 9, so multiply both sides of the equation by 9 written as a fraction $\left(\frac{9}{1}\right)$.

$$\frac{9}{1}\left(-\frac{2}{3} - \frac{4}{1}x\right) = \frac{9}{1}\left(\frac{5}{9}y\right)$$

$$-\frac{18}{3} - \frac{36}{1}x = \frac{45}{9}y$$

$$-6 - 36x = 5y$$

Move $5y$ to the left side of the equation by subtracting it from both sides. Also add 6 to both sides, so the constant ends up on the right side of the equation.

$$-36x - 5y = 6$$

The x-term has to be positive in order for the equation to be in standard form, but right now it's not. No sweat—just multiply both sides of the equation by -1.

$$36x + 5y = -6$$

CAUTION

Kelley's Cautions

Only the x-term has to be positive in standard form, not the y-term or the constant. As you can see in Example 4, changing the sign of the x-term (when necessary) will often result in one or more of the other terms becoming negative, and that's okay.

You've Got Problems

Problem 4: Put the linear equation $\frac{5}{4}y = \frac{7}{3} + \frac{1}{6}x$ in standard form.

Tricky Linear Equations

What do you do when you're trying to figure out the equation of a line but the slope isn't given to you? You still use the point-slope formula to write the equation—you just have to calculate the slope first. Basically, you'll be faced with two kinds of tricky linear equation tasks:

◆ Find the equation of a line given two points on that line

◆ Create a line either parallel or perpendicular to a second line

You'll need to use two different approaches to calculate slope in those two different types of problems.

How to Get from Point A to Point B

One classic type of algebra problem asks you to write the equation of a line that passes through two points. To solve those sorts of problems, you'll need the slope formula from Chapter 5, which states that a line passing through points (a,b) and (c,d) has the following slope:

$$m = \frac{d-b}{c-a}$$

Now it's time to put that formula to good use. Not only will you find the slope of the line passing through a couple of points, you'll find the equation of that line as well.

Example 5: Write the equation of the line that passes through the points (–4,7) and (2,–1) in standard form.

Solution: To create a linear equation, you need the slope of the line and one point on the line. This problem gives you *two* points on the line but no slope, so you need to use the slope formula.

$$m = \frac{-1-(7)}{2-(-4)}$$

$$m = \frac{-8}{6}$$

$$m = -\frac{4}{3}$$

Now that you have the slope, pick one of the given points and apply the point-slope formula. Here's what the substitution process looks like if you choose the point $(2,-1)$:

$$y - y_1 = m(x - x_1)$$

$$y - (-1) = -\frac{4}{3}(x - (2))$$

$$y + 1 = -\frac{4}{3}x + \frac{8}{3}$$

To write the line in standard form you'll need to eliminate those fractions, so multiply everything by 3.

$$3(y + 1) = \frac{3}{1}\left(-\frac{4}{3}x + \frac{8}{3}\right)$$

$$3y + 3 = -\frac{12}{3}x + \frac{24}{3}$$

$$3y + 3 = -4x + 8$$

Move the variables to the left side of the equation and the constants to the right side.

$$4x + 3y = 5$$

Critical Point

Once you figure out the slope in Example 5, it doesn't matter which point you plug into point-slope form—both give you the same final answer. Here's what the work looks like if you choose the point $(-4,7)$ instead of $(2,-1)$:

$$y - y_1 = m(x - x_1)$$

$$y - (7) = -\frac{4}{3}(x - (-4))$$

$$y - 7 = -\frac{4}{3}x - \frac{16}{3}$$

$$3(y - 7) = 3\left(-\frac{4}{3}x - \frac{16}{3}\right)$$

$$3y - 21 = -4x - 16$$

$$4x + 3y = 5$$

You can check your answer by plugging both of the original points into the equation. If they each result in true statements, you did everything right.

Plug in $(-4, 7)$	Plug in $(2, -1)$
$4(-4) + 3(7) = 5$	$4(2) + 3(-1) = 5$
$-16 + 21 = 5$	$8 - 3 = 5$
$5 = 5$	$5 = 5$

You've Got Problems

Problem 5: Write the equation of the line that passes through the points $(-3, 5)$ and $(-8, 0)$ in standard form.

Parallel and Perpendicular Lines

You may not know a lot about geometry yet, and that's okay. (By the way, the technical definition of geometry, according to *Webster's Dictionary*, is "What an acorn says when it grows up." Get it? "Gee-I'm-a-tree?") However, you may be asked to construct equations of parallel and perpendicular lines, so here's a quick explanation of those concepts:

♦ *Parallel lines* never intersect one another. Like the rails on a train track, they run side by side but never touch one another. That's because parallel lines have the same exact slope, which keeps them the same exact distance from each other forever and ever.

♦ *Perpendicular lines* intersect one another at right, or 90-degree, angles. You probably know what right angles are, but in case you don't, picture a square or rectangle—their sides meet at 90-degree angles. Perpendicular lines have slopes which are opposite reciprocals.

Figure 6.2 contains a set of parallel and perpendicular lines and indicates the slope of each line.

Talk the Talk

Parallel lines do not intersect because they have the same slope. **Perpendicular lines** meet at 90-degree angles, thanks to slopes which are opposite reciprocals. In other words, if lines l and n are perpendicular and the slope of line l is $\frac{a}{b}$, then the slope of line n must be $-\frac{b}{a}$.

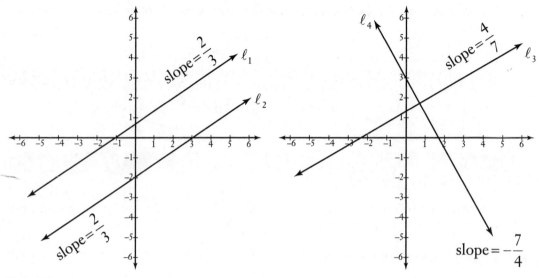

Figure 6.2

Parallel lines ℓ_1 and ℓ_2 have the same slope, whereas the slopes of perpendicular lines ℓ_3 and ℓ_4 are opposite reciprocals.

Example 6: Write the equation of line k in slope-intercept form if it passes through the point $(2,-3)$ and is perpendicular to the line $x - 5y = 7$.

Solution: If k is perpendicular to the line $x - 5y = 7$, their slopes are opposite reciprocals. Put $x - 5y = 7$ in slope-intercept form to figure out what its slope is. In other words, solve it for y:

$$-5y = -x + 7$$

$$y = \frac{1}{5}x - \frac{7}{5}$$

This line has slope $\frac{1}{5}$. Therefore, line k's slope is the opposite reciprocal: $-\frac{5}{1} = -5$. (Just flip the fraction upside down and multiply it by -1.)

Now that you have the slope of k and one of its points—the problem tells you that $(2,-3)$ is a point on line k—apply the point-slope formula.

$$y - y_1 = m(x - x_1)$$
$$y - (-3) = -5(x - (2))$$
$$y + 3 = -5x + 10$$

The problem asks for the answer in slope-intercept form, so solve for y.

$$y = -5x + 7$$

The Least You Need to Know

♦ The point-slope formula states that a line with slope m passing through the point (x_1, y_1) has equation $y - y_1 = m(x - x_1)$.

♦ To put an equation into slope-intercept form, solve it for y: $y = mx + b$. The slope of the line is m, and b is the y-intercept.

♦ Parallel lines have equal slopes and perpendicular lines have slopes that are opposite reciprocals.

Linear Inequalities

In This Chapter

◆ Differentiating between equations and inequalities

◆ Solving and graphing basic inequalities

◆ Interpreting compound inequalities

◆ Graphical solutions for linear inequalities

When Thomas Jefferson wrote that "all men are created equal" in the Declaration of Independence, he probably meant it in a philosophical and political way. That is to say, all people should have the same rights and responsibilities as citizens of the United States. Practically speaking, it would be hard to argue that all men (and women) are *actually* created equal. If that were true, it wouldn't make much difference if the coach of the St. Louis Rams decided to let me play quarterback during the playoffs.

If all men were created equal, I would possess immeasurable athletic talent, and could throw a 60-yard pass that resulted in the winning touchdown, to the praise and glory of thousands of fans, who would chant my name and buy my bobble-head dolls at the concession stand. In real life, if I were in for a single play, we both know what would happen. I'd get sacked by the meanest 350-pound defensive linemen you've ever seen, instantly breaking every major bone in my body, and single-handedly destroying any chance that my face would ever be attached to any toy, whether its head bobbled or not. (Unfortunately, my real head would probably bobble permanently from that point on, though the effect is less comical in real life, I assure you.)

Critical Point _____

Here's one way to remember which inequality sign is which: the *less* than symbol points *left*. Just remember "less goes left."

In any case, people are rarely created exactly equal, and so, too, mathematical statements are often unequal. In this chapter, I'll describe *inequalities*, which look a little bit like equations, but their sides aren't equal. Solving inequalities even feels a little like solving equations; the only *major* difference you'll find is that the graphs of inequalities look a bit different than the graphs of equations.

Equations vs. Inequalities

If the two sides of an inequality are not equal, there's got to be a reason why. The obvious explanation is that one side must be larger than the other. Knowing which side is larger is very important, so you'll use one of four inequality symbols (in place of an equal sign) to identify the larger side and to indicate whether or not there is even a remote possibility that the sides could be equal:

Kelley's Cautions _____

When it comes to negative numbers, the more negative a number is, the *less* it is (which almost seems backward). Therefore, $-17 < -9$ because both numbers are negative and 17 is larger than 9. For the same reason, you can write $-9 > -17$.

- $<$: This symbol means "less than." Use this to indicate that the quantity on the left side has a smaller value than the quantity on the right side. For example, the statement $-2 < 7$ literally means "negative 2 is less than 7."

- $>$: If you flip the less than symbol around, you get the "greater than" symbol, which means the exact opposite of the $<$ symbol. Therefore, you could write $100 > 57$, because 100 is greater than 57.

- $\leq$: This symbol means "less than or equal to"; basically, it means the same as the $<$ symbol, except that it also allows both sides to have the same value. In other words, the statement $6 \leq 10$ is true (because 6 is less than 10), but so is $7 \leq 7$ (because 7 is equal to itself).

♦ ≥: The "greater than or equal to" symbol is similar to the less than or equal to symbol, in that it also allows for the possibility of equality. Therefore, the statements $8 \geq -1$ and $-15 \geq -15$ are both true.

Critical Point

There is a fifth inequality symbol, ≠, which means "not equal to." You could use it to state that $10 \neq 4$ (10 is not equal to 4), which is true but not very helpful. It would be better to say $10 > 4$ (10 is greater than 4), because it explains *why* 10 and 4 aren't equal.

Example 1: Are the following statements true or false?

(a) $-7 > -1$

Solution: False; –7 is more negative, and therefore less than, –1.

(b) $1 \geq 1$

Solution: True; 1 is either greater than *or* equal to 1 (only one of those two conditions can be true at a time, and even though 1 is not greater than itself, it is equal to itself).

You've Got Problems

Problem 1: Are the following statements true or false?
      (a) $-3 < -3$
      (b) $5 \leq 11$

Solving Basic Inequalities

Most of the inequalities you'll see in algebra will contain variables. Although a statement like $1 < 7$ is easy enough to prove correct, it's really not all that intellectually stimulating or interesting. Instead, you'll be asked to solve an inequality statement like $-5x + 3 > -32$. Basically, your job will be to find the values of x that make that statement true. It's pretty much the same objective you had when solving equations, except in the world of inequalities, there are lots of solutions, not just one.

The Inequality Mood Swing

To solve an inequality that contains only one variable, follow the same steps you use to solve equations. In other words, simplify both sides of the inequality, isolate the variable, and then eliminate the variable's coefficient. However, there is one major difference between equations and inequalities: when solving an inequality, if you ever multiply or divide both sides by a negative number, you must reverse the inequality sign.

Critical Point _____

When you solve equations, you perform the same operations on both sides of the *equal sign*. Follow the same steps with inequalities, performing the operations on both sides of the *inequality sign*.

What do I mean by "reverse" the inequality sign? Change less than signs to greater than signs, and vice versa. (Less than or equal to signs become greater than or equal to signs, and vice versa.) I call this the inequality mood swing. Remember, it only happens when you multiply or divide by a negative number, and that only occurs when you're trying to eliminate the coefficient. So check to see if the coefficient is negative as you're eliminating it, and reverse the inequality sign as necessary.

Example 2: Solve the inequality $-5x + 3 > -32$.

Solution: Both sides of the inequality are already simplified (neither side contains like terms), so isolate the variable by subtracting 3 on both sides of the greater than sign. The inequality sign does not change here, because you're not *multiplying* or *dividing* by a negative number.

$$-5x > -35$$

Time to eliminate the coefficient—divide both sides by –5. Don't forget to reverse the inequality sign, because you're dividing by a negative number.

$$\frac{-5x}{-5} < \frac{-35}{-5}$$
$$x < 7$$

Any number less than 7, when plugged in for x, makes this inequality statement true. You obviously can't check them all to make sure your answer is right (you'd spend the rest of your life checking your work), but it doesn't hurt to check one answer just to make sure you're not way off. Here's how you could check the value $x = 6$, which should work, since it's less than 7:

$$-5x + 3 > -32$$
$$-5(6) + 3 > -32$$
$$-30 + 3 > -32$$
$$-27 > -32$$

Because –32 is more negative than –27, –32 < –27, and $x = 6$ is a valid solution to the inequality.

You've Got Problems

Problem 2: Solve the inequality $2(w - 6) \leq 18$.

Graphing Solutions

In Chapter 5, when I was discussing linear equations, I explained why it was important to draw graphs. When linear equations contain two variables, usually x and y, there are an infinite number of ordered pairs that made the equations true. Because you can't easily write an infinite list of solutions, it's handy to represent them with a graph.

You now know that basic inequalities also have an infinite number of solutions, so graphing will help you visualize those solutions as well. However, basic inequalities (like those from the previous section) are different than linear equations because they only contain one variable. Therefore, you don't need a coordinate plane to graph basic inequalities; all you need is a *number line*, pictured in Figure 7.1. Essentially, the number line is the x-axis from the coordinate plane; since there is no second variable, you don't need a second axis on the graph.

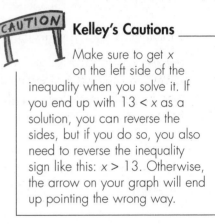

Kelley's Cautions

Make sure to get x on the left side of the inequality when you solve it. If you end up with $13 < x$ as a solution, you can reverse the sides, but if you do so, you also need to reverse the inequality sign like this: $x > 13$. Otherwise, the arrow on your graph will end up pointing the wrong way.

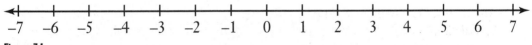

Figure 7.1

The number line is used to graph inequalities containing only one variable.

Here's how to graph the solution to a basic inequality:

1. **Solve the inequality.** Before you can draw the graph of an inequality, you need to know its solution.

Critical Point

A solid dot on the number line graph indicates that the given number should be included as a possible solution, whereas an open dot indicates that the value cannot be a solution. For example, the graph of $x > 7$ contains an open dot at 7 because it's not a valid answer (7 is not greater than itself).

2. **Draw a number line.** The number line doesn't always have to be centered at 0. It's sometimes more useful to center the number line at or near the value you found in the solution.

3. **Dot the hot spot.** When you solve the inequality, you'll get something like this: $x \le a$ or $x > a$, where a is a real number. If the inequality symbol allows for the possibility of equality ($\le$ or $\ge$), draw a solid dot at a. If, on the other hand, the symbol is either $<$ or $>$, draw an "open dot," or a circle, at the a value.

4. **It's not impolite to point.** Draw a dark arrow, starting at the dot, that points in the direction of the inequality symbol. If your final solution contains one of the less than symbols ($<$ or $\le$), draw the arrow to the left. If the solution contains a greater than symbol ($>$ or $\ge$), the arrow (like the symbol) points right.

Example 3: Graph the solution to $3(1 - 2x) < -5x + 6$.

Solution: First you need to solve the inequality for x. You want to isolate x on the left side of the inequality, so once you distribute the 3, add $5x$ to and subtract 3 from both sides.

$$
\begin{array}{rcl}
3 - 6x &<& -5x + 6 \\
-3 + 5x & & +5x - 3 \\
\hline
-x &<& 3
\end{array}
$$

Multiply (or divide) both sides by -1 to eliminate the negative sign in front of x, and don't forget to reverse the inequality sign.

$$x > -3$$

To graph this solution, draw an open dot (because the inequality symbol does not contain "or equal to") at -3 and sketch a dark arrow from that dot that extends to the right (because the inequality sign, like an arrowhead, points to the right), as illustrated by Figure 7.2.

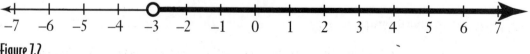

Figure 7.2

The graph of the inequality statement $x > -3$.

The graph makes it clear that every real number to the right of (but not including) –3 on the number line will make the inequality $x > -3$, and therefore the original inequality, $3(1 - 2x) < -5x + 6$, is true.

You've Got Problems
Problem 3: Graph the solution to $1 - x \geq 2x - 5$.

Compound Inequalities

You can define a range of values (like $x > 2$ and also $x \leq 9$) using a *single* inequality statement called a *compound inequality*, $2 < x \leq 9$. That statement reads "2 is less than x, which is less than or equal to 9." In other words, x has a value between 2 and 9—it might even be equal to 9, but it can't equal 2.

Talk the Talk

A **compound inequality** is a single statement used to represent two inequalities at once, such as $a < x < b$. It describes a range (or *interval*) of numbers with lower boundary a and upper boundary b. Whether or not those numbers are actually included in the interval is decided by the attached inequality sign. If the symbol is $\leq$, then the number is included; if it's $<$, then the boundary is not included.

Take a look at Figure 7.3, which contains the parts of a compound inequality. Notice that you should always write the lower boundary on the left and the upper boundary on the right. Furthermore, you should always use either the $<$ or $\leq$ symbol, or the statement may not make sense.

$$\text{lower boundary} \leq \text{variable} < \text{upper boundary}$$

Figure 7.3

The inequality symbols may be $<$, $\leq$, or any combination of the two.

Solving Compound Inequalities

A compound inequality has three pieces to it, rather than the usual two-sided equation or simple inequality. In order to solve a compound inequality, your goal will be t isolate the variable in the middle section. Do this by adding, subtracting, multiplying, and dividing the same thing to *all three parts* of the inequality at the same time.

CAUTION **Kelley's Cautions**

If you divide all three parts of a compound inequality by a negative number, reverse *both* inequality signs to get $b > x > a$, where b is the upper boundary and a is the lower boundary. I prefer $a < x < b$, which means the same thing but has its lower boundary on the left side.

Example 4: Solve the inequality $-4 \le 3x + 2 < 20$.

Solution: You want to isolate x where the $3x + 2$ currently is, so start by subtracting 2 from all three parts of the inequality.

$$-6 \le 3x < 18$$

To eliminate the x's coefficient, divide *everything* by 3.

$$-2 \le x < 6$$

So, any number between –2 and 6 (including –2 but excluding 6) will make the compound inequality true.

You've Got Problems

Problem 4: Solve the inequality $-1 < 2x + 5 < 13$.

Graphing Compound Inequalities

To graph a compound inequality, use dots to mark its endpoints on a number line. Just like in basic inequality graphs, the dots of compound inequalities correspond with the type of inequality symbol: $\le$ and $\ge$ symbols should be marked with a solid dot and $<$ and $>$ symbols should be marked with an open dot. A compound inequality statement has two inequality symbols and two endpoints, so use the symbol closest to each endpoint to decide what sort of dot to draw.

Once you've marked the boundaries, draw a dark segment connecting them. This dark segment indicates that all the numbers between the endpoints are solutions to the inequality.

Example 5: Graph the compound inequality $-14 < x \le -6$.

Solution: Because –14, the lower endpoint of this inequality, has a $<$ symbol next to it, draw an open dot at –14 on the number line. The upper boundary, on the other hand, has a $\le$ symbol next to it, so mark –6 with a solid dot. Finally, connect the two points with a dark line, as illustrated by Figure 7.4.

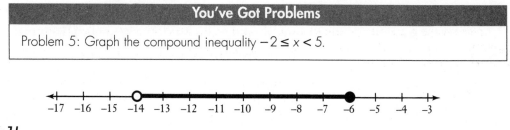

You've Got Problems

Problem 5: Graph the compound inequality $-2 \leq x < 5$.

Figure 7.4

The graph of $-14 < x \leq -6$. Note that this number line is not centered at 0. All the key points on the graph are negative, so it focuses on negative numbers instead.

Inequalities with Absolute Values

What is it about absolute values? Just when you get used to doing something one way, along come these little teeny bars that demand things be done their way. They are just like that high-maintenance boyfriend or girlfriend who wasn't content to let you live your life the way you always had. No, suddenly it was "Why can't we eat somewhere nice for a change?" and "Is there a federal law against you brushing your teeth more than once a month?" These adjustments aren't difficult to make, it's just that changing things once you get into a routine can be tricky.

Back in Chapter 5, you split absolute value equations into two distinct, nonabsolute value equations in order to reach a solution. Similarly, you have to break absolute value inequalities into two distinct, nonabsolute value inequalities to reach a solution. However, just to make things a little worse (if that were possible), the technique you use for inequalities with less than symbols is not the same one you use when you've got greater than symbols.

Inequalities Involving "Less Than"

If you're asked to solve an inequality problem containing absolute values, and the inequality symbol is either < or ≤, here are the steps you should follow to reach a solution:

1. **Isolate the absolute value expression on the left side of the inequality.**
   When you do, the problem should look something like this: $|x + a| < b$, where a and b are real numbers. (If a is negative, the problem will look like $|x - a| < b$.)

2. **Create a compound inequality.** Rewrite the statement $|x + a| < b$ as $-b < x + a < b$. In other words, drop the absolute value symbols, place a matching inequality symbol to the left of the statement, and then write the opposite of the constant to the left of that.

3. **Solve the compound inequality.** Use the procedures from the previous section to solve and/or graph the compound inequality.

Example 6: Solve the inequality $|2x - 1| + 3 \leq 6$ and graph the solution.

Solution: Subtract 3 from both sides of the inequality to isolate the absolute value expression.

$$|2x - 1| \leq 3$$

Drop the absolute value bars and write the opposite of 3 to the left of the expression. Between the newly added -3 and now "bar-free" expression, place a $\leq$ symbol to match the one already there.

$$-3 \leq 2x - 1 \leq 3$$

Solve the compound inequality by adding 1 to all three expressions and then dividing everything by 2.

$$-2 \leq 2x \leq 4$$
$$-1 \leq x \leq 2$$

Graph the solution by placing solid dots at -1 and 2 and connecting them, as illustrated by Figure 7.5.

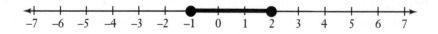

Figure 7.5

The graph of $|2x - 1| + 3 \leq 6$ matches the graph of the compound inequality $-1 \leq x \leq 2$.

If this problem had contained < instead of ≤ symbols, you would have solved it the exact same way. The graph would have looked very similar too, except the dots would have been open instead of closed.

You've Got Problems

Problem 6: Solve the inequality $4|x - 5| < 8$ and graph the solution.

Inequalities Involving "Greater Than"

Absolute value inequalities containing the symbols > or ≥ are solved much like their "less than" inequality sister problems. For one thing, you start by isolating the absolute value quantity first, and then you rewrite the inequality without absolute value bars. However, this time you won't end up with a compound inequality.

Rewrite the statement as two separate inequalities, one that looks just like the original (but doesn't contain absolute value bars), and the other with the inequality sign reversed and the opposite of the constant. That means $|x + a| > b$ becomes these two expressions:

$$x + a > b \qquad \text{or} \qquad x + a < -b$$

Note the word "or" between the expressions. That doesn't mean that you only have to write one *or* the other (they both need to be in your solution); it means that if you plug an x into *either* expression and it works for *just one* of them, then that x-value is a solution to the original inequality.

Example 7: Solve the inequality $|2x + 5| - 4 > -1$ and graph the solution.

Solution: Start by isolating the absolute value expression by adding 4 to both sides of the inequality.

$$|2x + 5| > 3$$

It's time to make those two inequality statements. Create the first simply by removing the bars: $2x + 5 > 3$. The second requires you to reverse the inequality sign and take the opposite of the constant to end up with $2x + 5 < -3$. (It's good form to write the word "or" between the statements.)

$$2x + 5 > 3 \qquad \text{or} \qquad 2x + 5 < -3$$

Solve the inequalities separately.

$$2x > -2 \qquad \text{or} \qquad 2x < -8$$
$$x > -1 \qquad \text{or} \qquad x < -4$$

That weird-looking two-headed monster is the answer. It says that any number greater than –1 *or* less than –4 will make the original inequality true. Because the solution consists of two inequality statements, the graph of the solution consists of both their graphs. Simply graph both of the inequalities on the same number line, as illustrated by Figure 7.6.

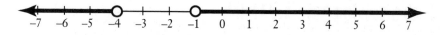

Figure 7.6

Create the solution graph of $|2x + 5| - 4 > -1$ *by graphing* $x > -1$ *and* $x < -4$ *on the same number line.*

You've Got Problems

Problem 7: Solve the inequality $|x - 4| \geq 2$ and graph the solution.

Graphing Linear Inequalities

Wouldn't it be great if you suddenly discovered that you had a skill you didn't know about? What if one day, while out for a jog, you found out that by twisting your right leg just a little, you could suddenly speak fluent Portuguese? (Assuming of course that you couldn't speak Portuguese before that.) That'd really be something to write home about, perhaps even in Portuguese. Well, you're about to find out that (with just a slight twist of your right leg) you can graph linear inequalities.

Critical Point

Equations and inequalities containing one variable are graphed on a number line. Equations and inequalities with two variables are graphed on a coordinate plane.

The major difference between basic inequalities and linear inequalities is that basic inequalities have one variable and linear inequalities have two, usually x and y. The handy thing about linear inequality graphs is that they are based on the graphs of linear equations, which are covered in Chapter 5. However, inequality graphs do have some characteristics that linear graphs do not:

- **The graph is not always a solid line.** If the inequality symbol in the statement is either < or >, then the line in the inequality graph will be dotted, instead of solid, to indicate that the points along the line are not solutions to the inequality. However, if the inequality symbol is ≤ or ≥, the line will be solid.

- **The solution is an entire region of the graph, not just the line.** The lines (whether solid or dotted) split the coordinate plane into two regions, like a fence would divide up a piece of property. All of the points on one side of the line will make the inequality true, whereas all the points on the other side will not. You indentify the region of solutions by lightly shading it on your graph.

To graph a linear inequality, treat it like a regular line and graph it. (Make sure to check whether the line should be solid or dotted, based on the inequality symbol.) Next, choose a coordinate, called a *test point*, from the coordinate plane and plug it into the original inequality. If it makes the inequality true, then all of the other points in the same region of the graph will as well, so shade that region. If the test point doesn't work, then the region on the *other* side of the line is the solution.

Critical Point _____

The dotted line in a linear inequality is the two-dimensional equivalent of the open circle in a basic inequality. Both mean "don't include this value (or coordinate pair) as a possible solution."

Example 8: Graph the inequality $2x - 3y < 12$.

Solution: For just a moment, pretend that the inequality symbol is an equal sign, and graph the resulting linear equation. The easiest way to graph a line is to calculate its intercepts (see Example 3 in Chapter 5 for an explanation). The inequality does not indicate "or equal to," so make the line dotted, as illustrated by Figure 7.7.

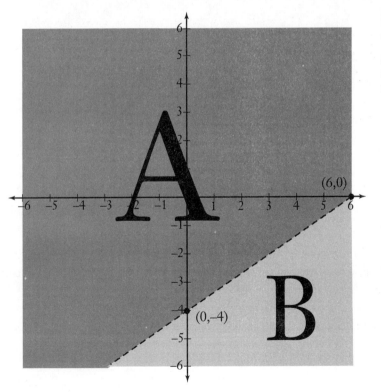

Figure 7.7

This is not yet the final graph, but notice how the line splits the coordinate plane into two regions, labeled A and B.

Choose a test point from one of the two regions. The origin, (0,0), is usually the best choice (unless the line passes through it), because what could be easier than plugging in 0 for both x and y? Whatever point you choose, plug in its x-coordinate for x in the inequality and its y-coordinate for y. Here's how it works for the origin:

$$2(0) - 3(0) < 12$$
$$0 < 12$$

Because this statement is true (0 is, indeed, less than 12), the origin is one solution of the inequality, and so is every other point in its region of the plane (region A in Figure 7.7). Shade in the solution region to complete the graph of the inequality, as illustrated by Figure 7.8.

Figure 7.8

The graph of 2x – 3y < 12 in all its splendor. Looks shady to me.

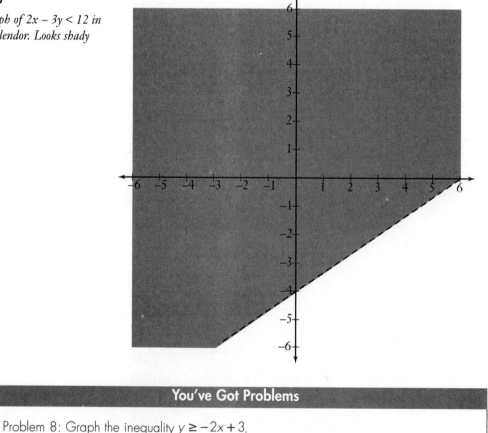

You've Got Problems

Problem 8: Graph the inequality $y \geq -2x + 3$.

The Least You Need to Know

- ◆ If you multiply or divide an inequality by a negative number, you must reverse the inequality sign.

- ◆ It is important to indicate (via open dots or dotted lines) points and lines that are not included in a graph because of the inequality symbols < and >.

- ◆ Compound inequalities express two inequalities in one statement.

- ◆ In order to solve inequalities containing absolute values, you must rewrite them as a pair of basic inequalities (or a single compound inequality) devoid of absolute values.

- ◆ Linear inequality graphs contain shaded solution regions.

Part 3

Systems of Equations and Matrix Algebra

You're about to solve bunches of equations, not one at a time, but simultaneously! (Forget mathematician! That sounds like the work of a mathemagician.) In this part, you use lots of different methods to solve groups of equations, including a foray into the land of the matrix. (Unfortunately, that's not as cool as it sounds.)

Chapter **8**

Systems of Linear Equations and Inequalities

In This Chapter

♦ Defining systems of equations

♦ Solving systems using different techniques

♦ Classifying systems with infinite or no solutions

♦ Graphing solutions to systems of inequalities

What could be more fun than one linear equation, you ask? (I know you didn't actually ask that, but play along.) Why, of course, the answer would be *two* linear equations at the same time (called a *system of equations*). Can you believe it? Two equations for the price of one, all for seven easy payments of $19.95—plus I'll throw in this contraption that lets you make your own beef jerky at home!

Talk the Talk

A system of equations is a collection of two or more equations. They're usually written with a left brace to indicate that they all belong to the same system:

$$\begin{cases} x + y = 7 \\ 2x - 1 = 4 \end{cases}$$

Your job will be to tell me what coordinate pair or coordinate pairs make *both* of the equations true. There are three basic ways to solve systems of equations: graphing, substitution, and elimination. As you explore them in this chapter, figure out which one you feel most comfortable with. However, keep in mind that some systems are easier to solve using one technique versus another, so make sure you can apply them all.

Solving a System by Graphing

The graph of a line is a visual representation of all the solutions to its linear equation. If you draw the graphs of two lines on a single coordinate plane, any point at which those graphs intersect represents a solution the linear equations have in common. Two lines usually intersect at a single point, so most systems of linear equations have a single solution.

In other words, to solve a system of equations all you have to do is graph the lines and figure out where they touch. The hard part, though, is making your graphs precise. Your answer is very dependent on how accurately you draw the graphs. Therefore, you should use graph paper and a ruler to make sure your measurements and lines are perfect.

Critical Point

Two lines usually intersect at a point, but sometimes it doesn't work like that. What if two lines are parallel (so they never intersect) or two lines completely overlap one another (so they intersect at every single point on their graphs)? You'll have either no solutions or more than you can count. I'll explain how to handle this later in the chapter.

Example 1: Solve the system of equations by graphing.

$$\begin{cases} x - 2y = 8 \\ -2x - y = -6 \end{cases}$$

Solution: Calculate the x- and y-intercepts of the lines and use those points to create your graphs (see Example 3 in Chapter 5 for more information). The equation $x - 2y = 8$ has intercepts (8,0) and (0,–4); the equation $-2x - y = -6$ has intercepts (3,0) and (0,6). As you can see in Figure 8.1, the graphs overlap at the point (4,–2).

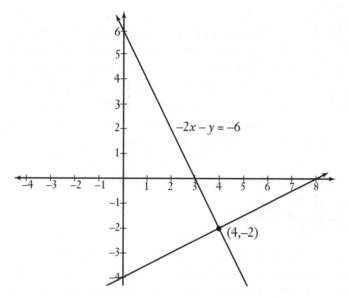

Figure 8.1

The solution to the system of equations is the point (4,–2), where the graphs of the equations intersect.

Actually, it would be more accurate to say that it *looks like* the graphs in Figure 8.1 overlap at (4,–2); after all, you'll have to draw these graphs by hand, so they won't be as precise as my computer-drawn graphs. To make sure the solution to the system actually is (4,–2), substitute $x = 4$ and $y = -2$ into both equations of the system. If you get true statements once you simplify, you can rest assured that (4,–2) doesn't just *look like* the intersection point of those lines—it definitely *is* the intersection point.

Critical Point

Either write the solution to the system in Example 1 as a point, (4,–2), or as a set of *x-* and *y*-coordinates: $x = 4$ and $y = 2$.

$$x - 2y = 8 \qquad -2x - y = -6$$
$$4 - 2(-2) = 8 \qquad -2(4) - (-2) = -6$$
$$4 + 4 = 8 \qquad -8 + 2 = -6$$
$$8 = 8 \qquad -6 = -6$$

Luckily, the coordinates of the solution were integers, which are easy to spot on the coordinate plane. However, if the solution had been something much uglier, like $\left(4\frac{1}{19}, -2\frac{3}{8}\right)$, there would have been no way to get that answer just by looking at the intersection point on a hand-drawn graph. Therefore, this method of solving systems of equations is not used very often. More sophisticated (and less subjective) methods, which you'll learn about in the next couple of sections, are necessary. Besides, who wants to carry graph paper around with them all the time?

You've Got Problems

Problem 1: Solve the system by graphing.

$$\begin{cases} 2x + y = 3 \\ x - 3y = -9 \end{cases}$$

The Substitution Method

If it's easy to isolate a variable in one of the equations of a system, then that system is a prime candidate for the substitution method. The great thing about this alternative to the graphing method is that you don't need any new skills—you just need to know how to solve an equation for a variable, which is covered in Chapter 4.

By the way, the substitution method can be used to solve *any* system of equations, and if the solution is a single point (it usually is), this technique will *always* get you the correct answer. However, the more complicated the equations in the system, the more arithmetic is involved, and the easier it is to make a mistake.

The best time to use the substitution method is when one of the variables in the system has a coefficient of 1 or –1. (In other words, there is no numeric coefficient written next to the variable.) That "absent" coefficient makes solving for the corresponding variable easier than falling down the stairs (which I know for a fact is a very easy thing to do, considering the amazing number of times I have managed to do it).

 Kelley's Cautions

Once you solve one of the equations in a system for a variable, make sure to plug the result into the *other* equation. If you plug the result back into the original form of the equation you solved, all the terms will cancel out, and you'll eventually get 0 = 0 (a true but uninteresting statement).

Example 2: Use the substitution method to solve the system.

$$\begin{cases} 7x + 4y = 6 \\ 2x - y = -9 \end{cases}$$

Solution: The coefficient of y in the second equation is –1, which makes solving for y pretty simple. Subtract $2x$ from both sides of that equation and multiply everything by –1.

$$-y = -2x - 9$$
$$-1(-y) = -1(-2x - 9)$$
$$y = 2x + 9$$

How'd They Do That?

The substitution method eliminates one of the variables in an equation. By solving the second equation for y in Example 2, you are actually rewriting y as a statement containing x's. So when you replace the y with x's in the other equation, you get an equation with all x's in it, which will have only one answer. This is much better than the original linear equation $2x - y = -9$, which (like all linear equations) has an infinite number of coordinate pair solutions.

Now that you know $y = 2x + 9$, replace y in the other equation with $2x + 9$.

$$7x + 4y = 6$$
$$7x + 4(2x + 9) = 6$$

Distribute 4 and solve for x.

$$7x + 8x + 36 = 6$$
$$15x + 36 = 6$$
$$15x = -30$$
$$x = -2$$

You've got half of the correct answer—all you need is the y-value to complete the coordinate pair. Plug $x = -2$ into the equation you solved for y.

$$y = 2x + 9$$
$$y = 2(-2) + 9$$
$$y = -4 + 9$$
$$y = 5$$

The solution to the system is $(-2,5)$, which you could also write as $x = -2, y = 5$.

You've Got Problems

Problem 2: Use the substitution method to solve the system.
$$\begin{cases} x - 4y = 11 \\ 3x + 7y = -5 \end{cases}$$

The Elimination Method

As the "How'd They Do That?" sidebar in the last section suggests, the substitution method is basically a sneaky way to eliminate a variable. It's the boxing equivalent of a sucker punch. Before the system knew what hit it, wham! It's over! One of the variables is lying unconscious in the boxing ring, and you can handle the variable that's left with no trouble. (After all, two against one isn't fair, is it?)

If the substitution method is a sucker punch, then solving systems of equations with the *elimination method* is the metaphorical equivalent of a full-blown wrestling match. There's no finesse or devious tactical plan at work here—you simply stride in, manhandle the equations with brute force, and show them you're in charge. Like the substitution method, elimination will work for any system that has a valid solution. It will even shed light on the situation when there isn't an answer, which is discussed in the next section.

As the name suggests, the whole purpose of the elimination method is to eliminate a variable, and you'll do that by multiplying one or both of the equations in the system by a real number. Multiplying by a real number, as long as you do it to both sides of the equation, won't change the equation's solution, so you may be wondering how that helps. Well, you don't just multiply by numbers that you pull out of thin air—there's a trick to it. You multiply so that once you line up the like terms in both equations and add the equations together, one of the variables vanishes.

Once that variable's gone, the wrestling match is basically over. All that remains, before you can pin your opponent, is to solve the equation and evaluate the missing variable.

Critical Point _____

If you're having trouble figuring out what to multiply each equation by to eliminate a variable, use this little trick. First, write the equations in standard form (if they're not already). You'll end up with something like this:

$$\begin{cases} ax + by = c \\ dx + ey = f \end{cases}$$

(Of course, a, b, c, d, e, and f will be real numbers.) To eliminate x, multiply the top equation by d and the bottom equation by $-a$.

For example, in Example 3, you could have multiplied the top equation by 6 and the bottom equation by -2, and you would have gotten the same final answer.

Example 3: Solve the system using the elimination method.

$$\begin{cases} 2x - y = 13 \\ 6x + 4y = 4 \end{cases}$$

Solution: If you multiply the top equation by –3, you'll get $-6x + 3y = -39$. Why would you want to do that? Well, you'll end up with an x-coefficient of –6, which is the opposite of the x-coefficient in the other equation. If you add the new version of the top equation to the bottom equation, the x-terms disappear.

$$
\begin{array}{rcr}
-6x + 3y &=& -39 \\
+6x + 4y &=& 4 \\
\hline
7y &=& -35
\end{array}
$$

The resulting equation is very simple to solve—divide both sides by 7 and you get $y = -5$. Now it's time to figure out what x-value completes the ordered pair. Plug $y = -5$ into *either* of the linear equations in the system; you'll get the correct answer no matter which one you pick.

$$6x + 4y = 4$$
$$6x + 4(-5) = 4$$
$$6x - 20 = 4$$
$$6x = 24$$
$$x = 4$$

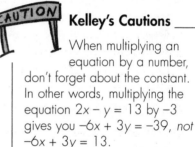

Kelley's Cautions

When multiplying an equation by a number, don't forget about the constant. In other words, multiplying the equation $2x - y = 13$ by –3 gives you $-6x + 3y = -39$, *not* $-6x + 3y = 13$.

The solution to this system is the coordinate pair (4,–5).

You've Got Problems

Problem 3: Solve the system using the elimination method.
$$\begin{cases} 2x - y = -11 \\ 5x + 2y = 4 \end{cases}$$

Systems That Are Out of Whack

Occasionally, as you attempt to solve a system of equations, something very bizarre will happen—when you open your algebra book, it will suck you in and transport you to the magical world of Narnia. Actually, that's not true. The weird happenstance is this: all of the variables disappear!

When all the variables vanish, it means one of two things has happened:

- **The system has no solutions.** If the statement you end up with is false, such as $0 = 7$, then the linear equations in the system have no common solutions, and the system is described as *inconsistent*. This happens when the graphs of the lines are parallel—there's no solution point because there's no intersection point.

Talk the Talk

A system of linear equations that has no solutions is **inconsistent**, whereas a system that has infinitely many solutions is **dependent**.

- **The system has infinitely many solutions.** If the statement you end up with is true, such as $5 = 5$, then the linear equations in the system have graphs that overlap, intersecting at every point on their graphs. Such systems are called *dependent*.

Example 4: Classify the system of equations.

$$\begin{cases} x - 2y = 1 \\ -3x + 6y = 3 \end{cases}$$

Solution: To solve this system via the elimination method, multiply the first equation by 3 and add it to the second equation.

$$
\begin{array}{rcrcr}
3x & - & 6y & = & 3 \\
-3x & + & 6y & = & 3 \\
\hline
0 & + & 0 & = & 6
\end{array}
$$

Hey wait a minute! The x- and y-terms vanished, too! All the variables are gone and the statement $0 = 6$ is false. That means the system is inconsistent and there are no solutions.

You get the same result with the substitution method. Solve the first equation for x: $x = 2y + 1$. Substitute that into the second equation, and you end up with a different (but still false) variable-free statement.

$$-3(2y+1)+6y=3$$
$$-6y-3+6y=3$$
$$-3=3$$

You've Got Problems

Problem 4: Classify the system of equations.
$$\begin{cases} 4x + 3y = -2 \\ -8x - 6y = 4 \end{cases}$$

Systems of Inequalities

According to Chapter 7, the graph of a linear inequality is a shaded region of the coordinate plane bounded by a line which is either solid or dotted, depending upon the inequality symbol involved. With that in mind, here's how to graph the solution to a *system* of inequalities:

1. **Graph each inequality of the system on the same coordinate plane.** Make sure to shade lightly, because when there are a bunch of inequalities the graph gets messy fast.

2. **The solution is the overlapping shaded area.** The region of the graph containing shading *from every inequality in the system* represents the solution.

The common shaded region is the solution to the system because it contains the coordinates that make *all* of the inequalities in the system true.

Example 5: Graph the solution to the system of inequalities.

$$\begin{cases} x > -3 \\ x - 2y \le 4 \end{cases}$$

Solution: Graph both inequalities on the same coordinate plane. (If you're wondering how to graph $x > -3$ because it has only one variable, treat it like the equation $x = -3$, the vertical line three units left of the y-axis.) You should end up with the graph in Figure 8.2.

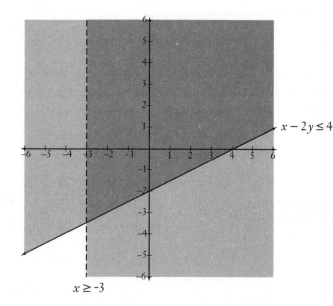

$x - 2y \le 4$

$x \ge -3$

Figure 8.2

Use test points to determine that you should shade to the right of the vertical line $x > -3$ and above the line $x - 2y \le 4$.

The most darkly shaded region in Figure 8.2 represents the solution of the system of inequalities. Some instructors will allow you to leave the graph as is, accepting the darkest region as the final answer, whereas some would rather you graph only the solution, as illustrated by Figure 8.3.

Figure 8.3

It's easier to identify the solution to the system of inequalities in this graph, compared to the graph in Figure 8.2.

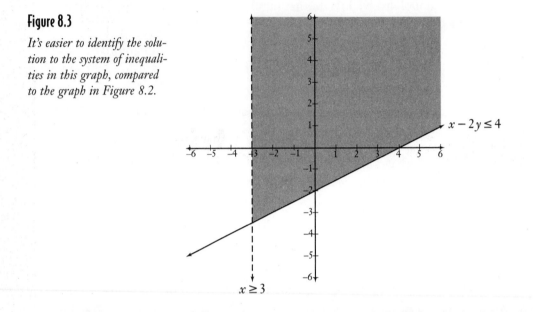

$x - 2y \leq 4$

$x \geq 3$

You've Got Problems

Problem 5: Graph the solution to the system of inequalities.

$$\begin{cases} y \leq -\dfrac{2}{3}x + 1 \\ y > 4x - 5 \end{cases}$$

The Least You Need to Know

♦ The solution to a system of equations makes all of the equations in the system true.

♦ A linear system of equations will have either one solution (a coordinate pair), no solutions (inconsistent systems), or infinitely many solutions (dependent systems).

♦ The elimination and substitution methods allow you to solve systems of equations by rewriting them in terms of one variable.

♦ The solutions to systems of linear inequalities are graphed as shaded regions on the coordinate plane.

The Basics of the Matrix

In This Chapter

- ◆ Defining a matrix and its component parts
- ◆ Adding, subtracting, and multiplying matrices
- ◆ Calculating determinants of matrices
- ◆ Solving systems of equations using determinants

When most people hear the word "matrix" nowadays, they think about people in dark suits who can jump from building to building while avoiding bullets in slow motion by contorting at impossible angles. Unfortunately for us all, there are a few differences between a mathematical matrix and a matrix as described by the science fiction movie.

For one thing, a math matrix contains almost no violence (none, in fact, unless you punch someone while trying to figure out your homework problems). Second of all, there are no attractive people skulking about in a math matrix. No beautiful women or strapping hunks wearing shiny black vinyl pants are fighting for the future of mankind—just a bunch of numbers written in rows and columns. In fact, the most attractive person in a math matrix is probably your ninth-grade math teacher, and the only pants he ever wore can be described most accurately as "brownish."

While the realm of the mathematical matrix may not be exciting, death-defying, or packed with stunts that cost millions of dollars, it is still pretty neat. They're completely different from the topics in Chapters 1 through 8 and the topics you'll see in the chapters that follow. As alien as they are, they have surprising uses. You'll even use them to solve systems of equations (in case the three techniques you learned in Chapter 8 didn't satisfy you).

What Is the Matrix?

Simply put, a *matrix* is a group of values called *elements* that are arranged in rows and columns and surrounded with brackets. The *order* of a matrix describes the number of rows and columns that matrix contains. Consider matrix A as defined below.

Talk the Talk

A **matrix** is a rectangular collection of values, called **elements,** arranged in rows and columns and surrounded by brackets. A matrix containing m rows and n columns is said to have **order** $m \times n$; if the matrix has the same number of rows and columns, it is described as **square.**

$$A = \begin{bmatrix} 2 & -1 & 3 & 0 & 9 \\ -8 & -6 & 4 & 7 & 13 \\ -20 & -3 & 8 & 11 & 1 \end{bmatrix}$$

Matrix A has order 3×5 (read "3 by 5"), because it contains three rows and five columns. (Rows are horizontal lines of numbers and columns are vertical lines of numbers.) When writing the order of a matrix, you always put the number of rows before the number of columns. If a matrix has an equal number of rows and columns, it is said to be a *square* matrix.

Critical Point

You can remember that *columns* are *vertical* because they both contain the letter c.

If you are discussing individual elements of a matrix, you use the notation a_{rc}, where r is the element's row, and c is its column. For example, in matrix A above, $a_{24} = 7$ because the number 7 appears in the second row and fourth column of the matrix. Notice that the lowercase variable used to represent the elements matches the uppercase variable of the matrix's name (a and A).

Matrix Operations

Although there are entire college courses dedicated to matrices and how useful they can be, in algebra you'll only be expected to perform a few matrix operations.

Multiplying by a Scalar

The simplest matrix operation requires you to multiply all of the numbers in the matrix by a separate real number, called a *scalar*. It's sort of like the distributive property on a grand scale. Multiply everything in the matrix by the scalar, and write the products in the same order as the original matrix.

Talk the Talk _____

A **scalar** is a real number. You use the term "scalar" to indicate that the number is not inside the matrix, as opposed to an "element," which is.

Example 1: If $A = \begin{bmatrix} -1 & 3 & 6 \\ -4 & 0 & -2 \end{bmatrix}$, evaluate $-2A$.

Solution: This problem asks you to multiply all of the elements in matrix A by the scalar value -2. It's a very simple task—just be careful with the signs as you multiply.

$$-2A = \begin{bmatrix} (-2)(-1) & (-2)(3) & (-2)(6) \\ (-2)(-4) & (-2)(0) & (-2)(-2) \end{bmatrix}$$

$$-2A = \begin{bmatrix} 2 & -6 & -12 \\ 8 & 0 & 4 \end{bmatrix}$$

You've Got Problems

Problem 1: If $B = \begin{bmatrix} 1 & 4 & 9 \\ -2 & 7 & -3 \\ 11 & -5 & 6 \end{bmatrix}$, evaluate $4B$.

Adding and Subtracting Matrices

In order to add two matrices, all you have to do is add up their corresponding elements. In other words, if you're asked to add matrices A and B together to get matrix C, start by adding the top left element in matrix A to the top left element of matrix B and write the result in the top left position of matrix C. Repeat this process for every pair of elements in corresponding spots.

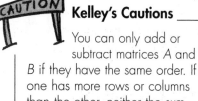

Kelley's Cautions _____

You can only add or subtract matrices A and B if they have the same order. If one has more rows or columns than the other, neither the sum nor the difference of the matrices is defined.

Subtracting matrices is no different. You just use the word "subtraction" if you happen to be multiplying one of the matrices by a negative scalar value, like in Example 2(b).

Example 2: If $A = \begin{bmatrix} 1 & 5 & -9 \\ 6 & -4 & 13 \\ 2 & -2 & 3 \end{bmatrix}$ and $B = \begin{bmatrix} 10 & 8 & -5 \\ 4 & 3 & -1 \\ 0 & 2 & 6 \end{bmatrix}$, evaluate the following expressions.

(a) $A + B$

Solution: Add each element of A to the corresponding element of B and write the result in the same position in the answer matrix.

$$A + B = \begin{bmatrix} 1+10 & 5+8 & -9+(-5) \\ 6+4 & -4+3 & 13+(-1) \\ 2+0 & -2+2 & 3+6 \end{bmatrix}$$

$$A + B = \begin{bmatrix} 11 & 13 & -14 \\ 10 & -1 & 12 \\ 2 & 0 & 9 \end{bmatrix}$$

(b) $2A - B$

Solution: In this expression, the elements of A are multiplied by a scalar value of 2, and the elements of B are multiplied by a scalar value of -1. Start by performing the scalar multiplication.

$$2A = \begin{bmatrix} 2 & 10 & -18 \\ 12 & -8 & 26 \\ 4 & -4 & 6 \end{bmatrix} \qquad -B = \begin{bmatrix} -10 & -8 & 5 \\ -4 & -3 & 1 \\ 0 & -2 & -6 \end{bmatrix}$$

Don't think to yourself, "I need to subtract matrices to get $2A - B$." Instead, think, "To get $2A - B$, I should *add* the matrices $2A$ and $-B$."

$$2A - B = \begin{bmatrix} 2 & 10 & -18 \\ 12 & -8 & 26 \\ 4 & -4 & 6 \end{bmatrix} + \begin{bmatrix} -10 & -8 & 5 \\ -4 & -3 & 1 \\ 0 & -2 & -6 \end{bmatrix}$$

$$2A - B = \begin{bmatrix} 2+(-10) & 10+(-8) & -18+5 \\ 12+(-4) & -8+(-3) & 26+1 \\ 4+0 & -4+(-2) & 6+(-6) \end{bmatrix}$$

$$2A - B = \begin{bmatrix} -8 & 2 & -13 \\ 8 & -11 & 27 \\ 4 & -6 & 0 \end{bmatrix}$$

Multiplying Matrices

The trickiest thing you'll need to do with matrices is to calculate their products. Unfortunately, matrix multiplication does not work like matrix addition—you can't just multiply the corresponding elements.

When Can You Multiply Matrices?

One of the biggest differences between matrix multiplication and matrix addition is that matrix multiplication does not require the two matrices involved to have the same order. Instead, multiplication has its own requirement: the number of *columns* in the first matrix must equal the number of *rows* in the second matrix. In other words, if matrix A has order $m \times n$ and matrix B has order $p \times q$, in order for the product $A \cdot B$ to exist, n and p must be equal.

Here's another little nugget: if the product of two matrices exists, it will have the same number of rows as the first matrix and the same number of columns as the second. In other words, if matrix A has order $m \times n$ and matrix B has order $n \times p$, $A \cdot B$ will have order $m \times p$.

Kelley's Cautions

Example 3 demonstrates a very important point—matrix multiplication is not commutative. Just because the matrix product A · B exists, there's no guarantee that B · A exists.

Example 3: If matrix C has order 4×5 and matrix D has order 5×9, does the product $C \cdot D$ exist? If so, describe the order of the product. Does the product $D \cdot C$ exist? If so, describe its order.

Solution: In order for a matrix product to exist, the number of *columns* of the left matrix must equal the number of *rows* of the right matrix. This is true for $C \cdot D$ (C has five columns and D has five rows), so the product exists. Furthermore, $C \cdot D$ will have order 4×9, the same number of rows as C and the same number of columns as D.

On the other hand, $D \cdot C$ does not exist, because matrix D has nine columns but matrix C only has four rows.

You've Got Problems

Problem 3: If A is a 3 x 4 matrix and B is a 3 x 3 matrix, which product exists: $A \cdot B$ or $B \cdot A$? What is the order of the product matrix?

Use Your Fingers to Calculate Matrix Products

If matrices A and B meet the row/column requirement and the product matrix P exists $(A \cdot B = P)$, quite a bit of work goes into calculating each element of P. The best way to learn how to multiply is via an example. All you'll need is a little patience, a couple of fingers, and the Ring of Power forged in the fiery bowels of Mordor (although the final requirement applies to you only if you are Frodo Baggins).

Example 4: Calculate the product.

$$\begin{bmatrix} 2 & -2 \\ 1 & 7 \end{bmatrix} \cdot \begin{bmatrix} -4 & 0 & -9 \\ 5 & 3 & -6 \end{bmatrix}$$

Solution: The first matrix is 2 x 2 and the second is 2 x 3. Because the number of columns in the left matrix matches the number of rows in the right matrix, the 2 x 3 product matrix exists. Start by creating a 2 x 3 matrix with generic labels for its elements.

$$\begin{bmatrix} p_{11} & p_{12} & p_{13} \\ p_{21} & p_{22} & p_{23} \end{bmatrix}$$

Each term is written in the form p_{rc}, where r indicates the element's row and c indicates its column.

Here's the key to multiplying matrices: each term p_{rc} is the result of multiplying the numbers in the rth row of the first matrix by the corresponding numbers in the cth column of the second matrix and then adding up all the results. Trust me, I know that sounds really weird and complicated, but it's not (or at least it won't be after I explain a little more).

The two little numbers after each p in the generic product matrix tell you where to put your fingers when you calculate the answer. The *left* little number tells you to put your *left* index finger on the *left*most number in that numbered row of the *left* matrix. The *right* little number tells you to put your *right* index finger on the *top* of that numbered column in the *right* matrix.

You'll need to do this for every p term in the product matrix, but as an example, I'll calculate the p_{13} term. The left little number is 1, so put your left index finger on the leftmost term in the *first* row of the left-hand matrix (the element 2 in this example). Since the right little number is 3, put your right index finger on the top number in the *third* column of the right-hand matrix (the element –9 in this example). The correct finger positions are shown in Figure 9.1.

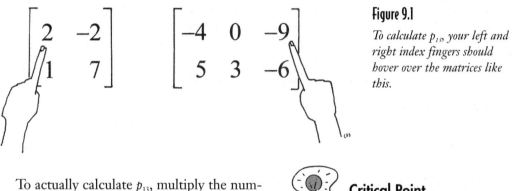

Figure 9.1

To calculate p_{13}, your left and right index fingers should hover over the matrices like this.

To actually calculate p_{13}, multiply the numbers you're pointing at: $2(-9) = -18$. Now move your left finger along the row to the next number (–2) and move your right finger along the column to the next number (–6) and multiply those numbers together: $(-2)(-6) = 12$. You continue doing this until you simultaneously reach the end of the row in the left matrix and the column in the right matrix (which is exactly what just happened), and then add up all the products you got along the way.

Critical Point

Now you know why the number of columns in the first matrix must match the number of rows in the second if you're going to multiply those matrices. It guarantees that your fingers will reach the end of a row and the bottom of a column simultaneously.

$$p_{13} = -18 + 12$$
$$p_{13} = -6$$

Here's the bad news: that's just one number finished in the product matrix! You'll have to follow the same process five more times to get the five other elements in the final product. Just to make sure you understand the finger-pointing process, here's how to get one of those five remaining values, the element p_{21}. Place your left index finger on element 1 (the leftmost number in row *two* of the left matrix) and your right index finger on element –4 (the topmost number in column *one* of the right matrix). Move your left finger horizontally and your right finger vertically (multiplying pairs of numbers as you go).

$$p_{21} = 1(-4) + 7(5)$$
$$p_{21} = -4 + 35$$
$$p_{21} = 31$$

When you calculate all of the elements of the product matrix at once, it looks like this:

$$\begin{bmatrix} 2 & -2 \\ 1 & 7 \end{bmatrix} \cdot \begin{bmatrix} -4 & 0 & -9 \\ 5 & 3 & -6 \end{bmatrix} = \begin{bmatrix} 2(-4)+(-2)(5) & 2(0)+(-2)(3) & 2(-9)+(-2)(-6) \\ 1(-4)+7(5) & 1(0)+7(3) & 1(-9)+7(-6) \end{bmatrix}$$

$$= \begin{bmatrix} -8-10 & 0-6 & -18+12 \\ -4+35 & 0+21 & -9-42 \end{bmatrix}$$

$$= \begin{bmatrix} -18 & -6 & -6 \\ 31 & 21 & -51 \end{bmatrix}$$

You've Got Problems

Problem 4: Calculate the product.

$$\begin{bmatrix} 2 & -5 \\ -3 & 1 \\ -4 & 0 \end{bmatrix} \cdot \begin{bmatrix} 6 & -2 \\ -1 & 3 \end{bmatrix}$$

Determining Determinants

A *determinant* is a real number value that is defined only for square matrices. At the moment, don't focus on *why* you find determinants—focus on *how* you find them. (The next section will explain one good use for the determinants, so suppress your curiosity until then if you can.) In addition, if you're getting frustrated, don't ask yourself, "Why should I learn algebra?"; instead, ask yourself, "How in the world did the opposite sex *ever* find me attractive until I decided to compute matrix products?" It's a better ego boost, and mathematicians use it all the time. Viva self-delusion!

Talk the Talk

Every square matrix has a real number associated with it called a **determinant;** you will use it to solve systems of linear equations later on in the chapter.

Even though all square matrices have determinants, in algebra, students focus on 2 x 2 and 3 x 3 matrices.

Once the dimensions get larger, the process becomes markedly more difficult, so it's usually discussed in more advanced courses like precalculus.

To indicate that you're calculating the determinant of a matrix, draw thin lines (which look a lot like, but definitely are not, absolute value bars) on both sides of the matrix, instead of brackets. For example, to indicate the determinant of matrix $A = \begin{bmatrix} 4 & 6 \\ -2 & -5 \end{bmatrix}$ you write $|A| = \begin{vmatrix} 4 & 6 \\ -2 & -5 \end{vmatrix}$.

2 × 2 Determinants

The determinant of the 2 × 2 matrix $\begin{vmatrix} a & b \\ c & d \end{vmatrix}$ is equal to $ad - cb$. In other words, multiply diagonally and then subtract the results, as illustrated in Figure 9.2.

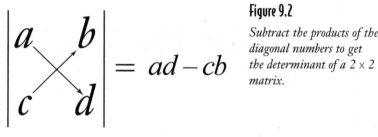

Figure 9.2

Subtract the products of the diagonal numbers to get the determinant of a 2 × 2 matrix.

Example 5: If $A = \begin{bmatrix} 4 & 6 \\ -2 & -5 \end{bmatrix}$, calculate $|A|$.

Solution: Picture the diagonal lines from Figure 9.2, which indicate that you should multiply $4(-5)$ and then subtract $-2(6)$. Don't forget that subtraction sign between $4(-5)$ and $-2(6)$ required by the formula. Even though $-2(6) = -12$ is a negative number, you still have to subtract it.

Kelley's Cautions

Make sure that you subtract in the correct order when calculating a 2 x 2 determinant, or you'll get the sign of your answer wrong. Always subtract the product of the upward diagonal.

$$|A| = 4(-5) - (-2)(6)$$
$$|A| = -20 - (-12)$$
$$|A| = -20 + 12$$
$$|A| = -8$$

3 × 3 Determinants

Even though calculating determinants of 3 × 3 matrices requires a bit more work, you do a lot of the same things you did with 2 × 2 determinants. In other words, you still multiply along diagonals and subtract those products; in fact, you subtract only the diagonals that point up and to the right, just like in 2 × 2 determinants. However, you have to rewrite a 3 × 3 matrix before you can do anything else.

Here's how to calculate the determinant of the 3 × 3 matrix $\begin{bmatrix} a & b & c \\ d & e & f \\ g & h & i \end{bmatrix}$:

1. **Copy the first two columns of the matrix.** Add a fourth and a fifth column to the matrix. The fourth column is a copy of the first column, and the fifth is a copy of the second.

$$\begin{bmatrix} a & b & c & a & b \\ d & e & f & d & e \\ g & h & i & g & h \end{bmatrix}$$

2. **Multiply along the diagonals.** In 2 × 2 determinants, you multiply diagonally down and then subtract the upward diagonal. Similarly, in 3 × 3 matrices, you multiply diagonally down three times (adding the results) and then subtract the three upward diagonals, as pictured in Figure 9.3.

Figure 9.3

Multiply along these diagonals to calculate the determinant of a 3 × 3 matrix. Make sure to subtract the product of each diagonal that points upward.

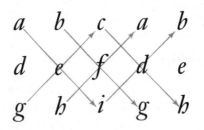

Multiply along the diagonals in Figure 9.3.

$$|A| = aei + bfg + cdh - gec - hfa - idb$$

Example 6: Calculate the determinant.

$$\begin{vmatrix} -3 & 5 & 1 \\ 2 & -6 & 0 \\ -7 & 4 & -1 \end{vmatrix}$$

Solution: Copy the first two columns to the right of the original matrix.

$$\begin{matrix} -3 & 5 & 1 & -3 & 5 \\ 2 & -6 & 0 & 2 & -6 \\ -7 & 4 & -1 & -7 & 4 \end{matrix}$$

Critical Point

Most graphing calculators can calculate determinants, which makes checking the answers to your homework and practice problems a breeze.

Multiply along the diagonals pictured in Figure 9.3, remembering to put a negative sign in front of every upward-pointing diagonal.

$$(-3)(-6)(-1) + (5)(0)(-7) + (1)(2)(4) - (-7)(-6)(1) - (4)(0)(3) - (-1)(2)(5)$$
$$= -18 + 0 + 8 - (42) - (0) - (-10)$$
$$= -18 + 8 - 42 + 10$$
$$= -42$$

You've Got Problems

Problem 6: Calculate the determinant.

$$\begin{vmatrix} -2 & -4 & 9 \\ 3 & -5 & 1 \\ 7 & 2 & -3 \end{vmatrix}$$

Cracking Cramer's Rule

Cramer's Rule is a neat trick that solves systems of equations, although I'd be lying if I called it a shortcut. Quite honestly, the elimination method from Chapter 8 solves systems faster, but Cramer's Rule is probably the easiest useful application of determinants, so that's why you find it in most algebra courses.

Most algebra students spend a lot of the year wondering "Why do I have to learn how to do any of this?" Cramer's Rule gives them the rare opportunity to see how learning a strange skill (like calculating determinants) is actually useful. True, solving systems of equations is not all that useful in day-to-day life ("Uh oh, my canoe just rolled over in the middle of the river—good thing I know how to solve systems of equations, or I'd never make it to shore!"), but at least it's something concrete in the maddeningly abstract world of algebra concepts.

Let's say you're trying to solve the system of equations

$$\begin{cases} ax + by = c \\ dx + ey = f \end{cases}$$

using Cramer's Rule. (Notice that both of the equations in the system are in standard form.) Here's your plan of attack:

1. **Write the coefficient matrix.** The coefficient matrix C is the 2×2 matrix containing the x-coefficients of the system in the first column and the y-coefficients in the second column.

$$C = \begin{bmatrix} a & b \\ d & e \end{bmatrix}$$

2. **Replace the columns in C one at a time.** Design two new 2×2 matrices, A and B, both based on the coefficient matrix C. To get matrix A, start with the coefficient matrix and replace the first column with the constants from the system (the numbers c and f from the right sides of the equal signs). In other words, instead of a and d in the first column of A, you'll have c and f. To get matrix B, replace the *second* column of the coefficient matrix with the constants.

$$A = \begin{bmatrix} c & b \\ f & e \end{bmatrix} \qquad B = \begin{bmatrix} a & c \\ d & f \end{bmatrix}$$

CAUTION Kelley's Cautions

If the system is inconsistent or dependent, $|C|$ will equal 0, meaning that Cramer's Rule will not work. (You wind up with $x = \dfrac{|A|}{|C|} = \dfrac{|A|}{0}$ and $y = \dfrac{|B|}{|C|} = \dfrac{|B|}{0}$, which are undefined.)

3. **Calculate the determinants of the matrices.** To figure out the solution to the system, you'll need to know the values of $|A|$, $|B|$, and $|C|$.

4. **Use the Cramer's Rule formulas.** The solution to the system (if it exists) is the coordinate pair (x,y), where $x = \dfrac{|A|}{|C|}$ and $y = \dfrac{|B|}{|C|}$.

That's a lot of steps, but most of them are very quick and painless, which will make Cramer's Rule a viable option for solving systems once you've practiced a few times and can fly through the steps quickly.

Example 7: Solve the system using Cramer's Rule.

$$\begin{cases} 4x + 6y = -9 \\ 3x - 8y = -13 \end{cases}$$

Solution: Create the coefficient matrix C, listing the x-coefficients in the first column and the y-coefficients in the second.

$$C = \begin{bmatrix} 4 & 6 \\ 3 & -8 \end{bmatrix}$$

How'd They Do That?

You can also apply Cramer's Rule to systems of three equations in three variables. Let's say you're solving this system via Cramer's Rule:

$$\begin{cases} ax + by + cz = d \\ ex + fy + gz = h \\ ix + ny + pz = q \end{cases}$$

The coefficient matrix C will be 3 x 3—three different variables in the system require three columns in the coefficient matrix, representing the coefficients of the x's, y's, and z's, respectively.

$$C = \begin{bmatrix} a & b & c \\ e & f & g \\ i & n & p \end{bmatrix}$$

Create matrix A by replacing the first column of C with the constants of the system, and create matrix B by sticking those constants into the second column (just like Cramer's Rule with 2 x 2 matrices). However, this system requires a fourth matrix D, in which the third column of the coefficient matrix is replaced by the constants.

$$A = \begin{bmatrix} d & b & c \\ h & f & g \\ q & n & p \end{bmatrix} \qquad B = \begin{bmatrix} a & d & c \\ e & h & g \\ i & q & p \end{bmatrix} \qquad D = \begin{bmatrix} a & b & d \\ e & f & h \\ i & n & q \end{bmatrix}$$

The solution to the system is:

$$x = \frac{|A|}{|C|}, \quad y = \frac{|B|}{|C|}, \quad z = \frac{|D|}{|C|}$$

To create matrix A, replace the first column of C with the column of constants (the numbers –9 and –13). To create matrix B, use the constants to replace the second column of matrix C.

$$A = \begin{bmatrix} -9 & 6 \\ -13 & -8 \end{bmatrix} \qquad B = \begin{bmatrix} 4 & -9 \\ 3 & -13 \end{bmatrix}$$

Next, calculate the determinants of all the matrices.

$$|A| = (-9)(-8) - (-13)(6) = 72 + 78 = 150$$
$$|B| = (4)(-13) - (3)(-9) = -52 + 27 = -25$$
$$|C| = (4)(-8) - (3)(6) = -32 - 18 = -50$$

You're nearly finished. Plug the determinants into the Cramer's Rule formula.

$$x = \frac{|A|}{|C|} = \frac{150}{-50} = -3$$

$$y = \frac{|B|}{|C|} = \frac{-25}{-50} = \frac{1}{2}$$

You've Got Problems

Problem 7: Solve the system using Cramer's Rule.
$$\begin{cases} 6x - 5y = -7 \\ 3x + 2y = 1 \end{cases}$$

The Least You Need to Know

♦ Matrices are rectangular collections of numbers, organized in rows and columns.

♦ In order for the matrix product $A \cdot B$ to exist, the number of columns in A must equal the number of rows in B.

♦ Determinants are defined for all square matrices.

♦ Cramer's Rule is a technique that solves systems of equations using matrices.

Part 4

Now You're Playing with (Exponential) Power!

It's time for an amicable breakup from linear equations. You should meet together at a local restaurant (so they won't cause a scene) and give them the old, "It's not you, it's me" speech. You need more out of life than slopes and intercepts. You need a graph that's a little more unpredictable, a graph with a little curve to it. In this part, you enter a strange new world of equations whose variables contain exponents. Sometimes it will feel a little strange and you'll yearn for the days of your first linear love, but trust me, this is best for both of you.

Chapter 10

Introducing Polynomials

In This Chapter

♦ Classifying polynomials

♦ Adding, subtracting, and multiplying polynomials

♦ Calculating polynomial quotients using long division

♦ Performing synthetic division

It's time to upgrade some of the algebra skills from the beginning of the book. Chapter 3 discussed basic exponential laws, which taught you that the product $(2x^4)(3x^7)$ is equal to $2 \cdot 3 \cdot x^{4+7} = 6x^{11}$. In Chapter 4, you may have manipulated the equation $3x = 9 + 2x$ by subtracting $2x$ from both sides to get $x = 9$. What you may not have known was that in both cases, you were actually simplifying polynomial expressions.

I haven't really discussed polynomials or gone into a lot of detail about simplifying them so far, mainly because the sorts of things you've done in the preceding chapters make sense intuitively—they sort of "feel right." Sure, $5x + 9x$ should equal $14x$; that seems to make sense, right? Before you go any farther, however, it's time to tie up any loose ends and get specific about what kinds of things you can and can't add together, and even explore some complicated topics, like dividing variable expressions.

Classifying Polynomials

A *polynomial* is basically a string of mathematical clumps (called *terms*) all added together. Each individual clump usually consists of one or more variables raised to exponential powers, often with a coefficient attached. Polynomials can be as simple as the expression $4x$, or as complicated as the expression $4x^3 + 3x^2 - 9x + 6$.

Polynomials are usually written in standard form, which means that the terms are listed in order from terms with the largest exponential value to the term with the smallest exponent. Because the term containing the variable raised to the highest power is listed first in standard form, its coefficient is called the *leading coefficient*. A term not containing a variable is called the *constant*.

Talk the Talk

A **polynomial** is a sum of distinct algebraic clumps (called **terms**), each of which consists of numbers and/or variables raised to exponential powers. The largest exponent in the polynomial is called the **degree**, and the coefficient of the variable raised to that exponent is called the **leading coefficient**. The **constant** in a polynomial is the term with no variable.

For example, if you were to write the polynomial $2x^3 - 7x^5 + 8x + 1$ in standard form, it would look like this: $-7x^5 + 2x^3 + 8x + 1$. (Note that each term's variable has a lower power than the term to its immediate left.) The *degree* of this polynomial is 5, its leading coefficient is -7, and the constant in the polynomial is 1.

Technically, a constant *does* have a variable attached to it, but the variable is raised to the 0 power. For example, you could rewrite the simple polynomial $2x + 1$ as $2x + 1x^0$, but since $x^0 = 1$ (and anything multiplied by 1 equals itself), there's no reason to write x^0 at the end of the polynomial.

Because there are so many different kinds of polynomials (52 flavors at last check, including pistachio), there are two techniques that are used to classify them, one based on the number of terms a polynomial contains (see Table 10.1), and one based on the degree of the polynomial (see Table 10.2).

Table 10.1–Classifying a Polynomial Based on the Number of Its Terms

Number of Terms	Classification	Example
1	monomial	$19x^2$
2	binomial	$3x^3 - 7x^2$
Number of Terms	Classification	Example
3	trinomial	$2x^2 + 5x - 1$

Notice that Table 10.1 only classifies polynomials with three or less terms. Polynomials with four or more terms are either classified according to degree (as demonstrated by Table 10.2) or merely described with the ultra-generic (and not very helpful) label "polynomial." (It's just as specific as labeling you a "human being.")

Table 10.2–Classifying a Polynomial Based on Its Degree

Degree	Classification	Example
0	constant	$2x^0$ or 2
1	linear	$6x^1 + 9$ or $6x + 9$
2	quadratic	$4x^2 - 25x + 6$
3	cubic	$x^3 - 1$
4	quartic	$2x^4 - 3x^2 + x - 8$
5	quintic	$3x^5 - 7x^3 - 2$

There are more degree classifications for polynomials, but those listed in Table 10.2 are by far the most commonly used.

When classifying a polynomial, you don't have to choose one method or the other. Actually, using both whenever possible paints a more descriptive picture.

Example 1: Classify the following polynomials.

(a) $3 - 4x - 6x^2$

Critical Point

If you're asked to classify a polynomial like $3x^3y^2 - 4xy^3 + 6x$ (which contains more than one kind of variable in some or all of its terms) according to its degree, add the exponents in each term together. The highest total is the degree. The degree of $3x^3y^2 - 4xy^3 + 6x$ is 5, because the highest exponent total comes from the first term: $3 + 2 = 5$.

Solution: This polynomial has three terms, so it's a trinomial. Furthermore, its degree is 2, which makes it quadratic. Therefore, it's a quadratic trinomial. When you use both classifications at once, write the degree classifier first because it's an adjective ("trinomial quadratic" just doesn't sound right).

(b) 13

Solution: There's only one term and it has no variable written explicitly. Therefore, the terms 13 and $13x^0$ mean the same thing, and both are constant monomials.

You've Got Problems

Problem 1: Classify the following polynomials:

(a) $4x^3 + 2$

(b) $11x$

Adding and Subtracting Polynomials

In previous chapters, you've simplified expressions like $3x + 7x$ to get $10x$, or perhaps subtracted terms like $5y - 9y$ to get $-4y$. That arithmetic makes perfect sense if you translate the mathematics into words. For example, the expression $3x + 7x$ literally means "three of a certain number added to seven more of that same number," which equals "10 of that number," or $10x$.

Talk the Talk

Like terms have variables which match exactly, like $4x^2y^3$ and $-7x^2y^3$. You can only add or subtract two terms if they are like terms.

Kelley's Cautions

Many students try to simplify the expression $4x + 5y$ as $9xy$, but that's wrong! You can't add or subtract $4x$ and $5y$ because the terms have different variables. It would be like adding four cats to five dogs and getting nine dats (or cogs). Unlike terms are apples and oranges—you can't combine them.

You are only allowed to combine the coefficients of those terms because they contain *the exact same variables*. Any two terms whose variables match *exactly* are called *like terms*. Here's the reason you care: you cannot add or subtract terms unless they are like terms.

If two terms have the same variables and get all nervous when they look at each other, you can upgrade them from *like* terms to *love* terms, but it's hard to read the emotions of variables (they're always changing on you), so most mathematicians don't bother trying to make that distinction.

Therefore, the expression $13x^2y^3 - 5x^2y^3$ can be simplified as $8x^2y^3$. The variables in both terms match exactly (they both contain x^2y^3), so all you have to do is combine the coefficients (13 − 5 = 8) and attach a copy of the matching variables.

Example 2: Simplify the following expression.

$$4x^3 + 5x^2 - 3x + 1 - (2x^3 - 8x^2 + 9x - 6)$$

Solution: Start by applying the distributive property—multiply everything in the parentheses by −1.

$$4x^3 + 5x^2 - 3x + 1 - 2x^3 + 8x^2 - 9x + 6$$

Rewrite the expression so that all like terms are grouped together; it makes simplifying easier.

$$4x^3 - 2x^3 + 5x^2 + 8x^2 - 3x - 9x + 1 + 6$$

Combine like terms.

$$(4-2)x^3 + (5+8)x^2 + (-3-9)x + (1+6)$$
$$= 2x^3 + 13x^2 - 12x + 7$$

You've Got Problems

Problem 2: Simplify the expression $3x^2 - 9 + 2(x^2 - 6x + 5)$.

Multiplying Polynomials

Unlike addition and subtraction, you don't need like terms in order to multiply polynomials (nor do you need like terms to divide polynomials—more on that in the next section). That makes multiplying polynomials pretty easy. All you have to do is apply the exponential rules and the distributive property from Chapter 3.

Products of Monomials

Here's what you should do to multiply two monomials together:

1. **Multiply their coefficients.** The result is the coefficient of the answer.

2. **List all the variables that appear in either term.** Write these variables to the left of the coefficient you got in step 1, preferably in alphabetical order.

How'd They Do That?

You add the powers of matching variables in the third step because of the exponential rule $x^a \cdot x^b = x^{a+b}$. (The product of exponential expressions with matching bases equals the common base raised to the sum of the powers.)

3. **Add up the powers.** Total the exponents for each variable individually and write those totals above the variables in the answer.

Even if the steps seem weird at first, don't worry. Multiplying monomials is a skill you'll pick up very quickly.

Example 3: Calculate the following products.

(a) $(-3x^2y^3z^5)(7xz^3)$

Solution: Start by multiplying the coefficients: $-3 \cdot 7 = -21$. Next, list all the variables that appear in the problem in alphabetical order. It doesn't matter that the second monomial doesn't contain y. As long as a variable appears *anywhere* in the problem, you should write it next to the coefficient you just calculated.

$$-21xyz$$

Add up the exponents for each variable listed. The expression $-3x^2y^3z^5$ contains x to the second power, and $7xz^3$ contains x to the first power. Therefore, their product will contain x to the $2 + 1 = 3$ power. Similarly, the z power of the answer should be 8, since the monomials contain z^5 and z^3. There's only one y-term (y^3), so just copy its power to the final answer.

$$-21x^3y^3z^8$$

(b) $3w^2x(2wxy - x^2y^2)$

Solution: Apply the distributive property, multiplying both terms by $3w^2x$.

$$3w^2x(2wxy) + 3w^2x(-x^2y^2)$$

Calculate each product separately.

$$3 \cdot 2 \cdot w^{2+1} \cdot x^{1+1} \cdot y + 3(-1) \cdot w^2 \cdot x^{1+2} \cdot y^2$$
$$= 6w^3x^2y - 3w^2x^3y^2$$

You've Got Problems

Problem 3: Calculate the product $3x^2y(5x^3 + 4x^2y - 2y^5)$.

Binomials, Trinomials, and Beyond

Because multiplying polynomials doesn't require like terms, it's sort of liberating. However, the examples so far have only dealt with multiplying monomials. Happily, multiplying polynomials with multiple terms is not very difficult either—you just use a slightly modified version of the distributive property.

Thanks to the distributive property, you know that $a(b + c)$ is equal to $ab + ac$. Distributing a means you're multiplying each term in the parentheses by a. In a similar fashion, you can multiply the binomials $a + b$ and $c + d$. Instead of distributing a single monomial, you distribute each term in the first binomial to each term in the second binomial, one at a time. In other words, multiply both terms of $c + d$ by a and then go back and multiply both terms by b.

$$ac + ad + bc + bd$$

Critical Point _____

Some algebra teachers focus on the FOIL method, a technique for multiplying two binomials. Each letter stands for a pair of terms in the binomials—the first, outside, inside, and last terms.

If you've never heard of FOIL, that's fine, because it only works when you're multiplying two binomials, whereas my multiple distribution technique works for all polynomial products. Besides, if you use my method, you end up getting the same results as the FOIL method when you multiply binomials.

So you're still distributing—you're just doing it twice. What if you want to multiply a trinomial by a trinomial? Follow the same procedure, distributing each term of the first polynomial to each term of the second, one at a time.

$$(a + b + c)(d + e + f) = ad + ae + af + bd + be + bf + cd + ce + cf$$

In case you're wondering, the numbers of terms in the polynomials you're multiplying don't have to match. You can multiply a binomial by a trinomial just as easily, as you'll see in Example 4.

Example 4: Expand and simplify the expression $(x - 2y)(x^2 + 2xy - y^2)$.

Solution: Both terms of the left polynomial (x and $-2y$) should be distributed through the second polynomial, one at a time.

$$(x)(x^2)+(x)(2xy)+(x)(-y^2)+(-2y)(x^2)+(-2y)(2xy)+(-2y)(-y^2)$$
$$=x^{2+1}+2x^{1+1}y-xy^2-2x^2y-2\cdot 2xy^{1+1}+2y^{1+2}$$
$$=x^3+2x^2y-xy^2-2x^2y-4xy^2+2y^3$$

Kelley's Cautions

After you multiply polynomials, always check to see if you can simplify the result. All algebra teachers demand simplified answers, and if you don't comply, they've been known to do things like mark answers wrong, take points off, or (in extreme cases) get so angry that they send a cybernetic organism back in time to kill you before you can sign up for their class.

Now combine like terms. If you look closely, you'll see that $2x^2y$ and $-2x^2y$ have the same variable, so they can be combined to get 0 (they're opposites of one another and therefore cancel each other out). Also combine the terms $-xy^2$ and $-4xy^2$ to get $-5xy^2$.

$$x^3 - 5xy^2 + 2y^3$$

You've Got Problems

Problem 4: Expand and simplify the expression $(2x + y)(x - 3y)$.

Dividing Polynomials

There are two techniques you can use to divide polynomials. One (which may feel a bit familiar) will work for all polynomial division problems but takes a while, whereas the other is much faster but only works in specific circumstances.

Long Division

Even though it's a bit cumbersome, the most reliable way to divide polynomials is with old-fashioned long division. It works exactly the same as the technique you learned in elementary school to divide whole numbers, so it may feel familiar.

Example 5: Calculate the quotient $(x^3 + 5x^2 - 3x + 4) \div (x^2 + 1)$.

Solution: Start by rewriting the problem in long division notation. The *divisor* (what you're dividing by) goes to the left, and the *dividend* (what you're dividing into) goes beneath the symbol. As you rewrite the polynomials, make sure there are no missing exponents in either the divisor or the divi-dend.

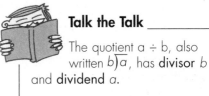

Talk the Talk _____

The quotient $a \div b$, also written $b\overline{)a}$, has **divisor** b and **dividend** a.

In this problem, the divisor $x^2 + 1$ has no x-term. Write the missing x-term in there with a coefficient of 0—if you don't, things won't line up right.

$$x^2 + 0x + 1\overline{)x^3 + 5x^2 - 3x + 4}$$

Look at the first term of the divisor, x^2, and the first term of the dividend, x^3. Ask yourself, "What times x^2 will give me *exactly* x^3?" The answer is x, so write that above the division symbol, directly above $-3x$ because it and the number you just came up with, x, are like terms.

$$\begin{array}{r} x \\ x^2 + 0x + 1\overline{)x^3 + 5x^2 - 3x + 4} \end{array}$$

Distribute that x to each term of the divisor, and write the result below the dividend so that the like terms line up; draw a horizontal line beneath the product.

$$\begin{array}{r} x \\ x^2 + 0x + 1\overline{)x^3 + 5x^2 - 3x + 4} \\ \underline{x^3 + 0x^2 + x } \end{array}$$

Multiply everything in that bottom line by –1 and then combine the like terms stacked on top of each other. Write the result below the hori-zontal line.

$$\begin{array}{r} x \\ x^2 + 0x + 1\overline{)x^3 + 5x^2 - 3x + 4} \\ \underline{-x^3 - 0x^2 - x } \\ 5x^2 - 4x \end{array}$$

How'd They Do That?

You ask yourself "What times x^2 will give me *exactly* x^3?" because you're trying to eliminate the first term of the dividend.

Copy the final term of the dividend below the horizontal line, next to $5x^2 - 4x$. Repeat the process, beginning with the question, "What times x^2 (the first term in the divisor) will give me exactly $5x^2$?" (the first term in the subtraction problem you just finished); the answer is 5. Write that constant above the division symbol (above 4, its like term), distribute it to each term of the divisor, change each of the terms in the product to its opposite, and combine like terms.

$$
\begin{array}{r}
x+5 \\
x^2+0x+1 \overline{)x^3+5x^2-3x+4} \\
-x^3-0x^2-x \\
\hline
5x^2-4x+4 \\
-5x^2-0x-5 \\
\hline
-4x-1
\end{array}
$$

If there had been additional terms in the dividend, you'd drop each term down one at a time and repeat the whole process again, but since you just dropped the constant 4 (the last term of the dividend), you're finished. The quotient is the quantity above the division symbol ($x + 5$) and the remainder is $-4x - 1$.

Write the answer as the quotient plus a fraction whose numerator is the remainder and whose denominator is the original divisor.

$$x+5+\frac{-4x-1}{x^2+1}$$

Check your answer by multiplying the quotient $x + 5$ and the original divisor $x^2 + 1$ and then adding the remainder.

$$
\begin{aligned}
&(\text{quotient})(\text{divisor})+\text{remainder} \\
&=(x+5)(x^2+1)+(-4x-1) \\
&=x^3+x+5x^2+5-4x-1 \\
&=x^3+5x^2+x-4x+5-1 \\
&=x^3+5x^2-3x+4
\end{aligned}
$$

If you did everything correctly, you'll get the original dividend. That's exactly what happened here, so bask in the glow of your own mathematical greatness, secure in the knowledge that you rule.

You've Got Problems

Problem 5: Calculate the quotient $(x^2 - 7x + 8) \div (x + 4)$.

Synthetic Division

When dividing by a linear binomial (like $x + 2$ or $x - 5$), *synthetic division* is the fastest way to the answer. For instance, the division problem $(x^3 - 2x^2 + 3x - 4) \div (x + 3)$ is an ideal candidate for synthetic division, but the problem $(x^3 - 2x^2 + 3x - 4) \div (x^2 + 3)$ is not, because the divisor is not linear. In a nutshell, it works when you divide by "$x +$ a number" or "$x -$ a number"

Synthetic division is much simpler than long division. A few seconds of playing with some coefficients is all it takes to calculate a quotient. I don't know why it's called "synthetic" division—it's not unnatural, filled with preservatives, or fake. It's never been to the plastic surgeon to get a nip, a tuck, an implant, or a reduction, so the name puzzles me. Like long division, the best way to learn the process is through an example, so here goes nothing.

Talk the Talk

Synthetic division is a quick way to calculate polynomial quotients when the divisor is a linear binomial.

Example 6: Calculate the quotient $(2x^3 - x + 4) \div (x + 3)$.

Solution: Check to see if there are any missing powers of x in the dividend. Hmmm, there's no x^2-term in $2x^3 - x + 4$, so insert it with a coefficient of 0 (just like you do in long division) to get a dividend of $2x^3 + 0x^2 - x + 4$. List those coefficients in order, starting with the highest exponent and working to the lowest.

$$2 \quad 0 \quad -1 \quad 4$$

To the left of that list, write the *opposite* of the constant from the divisor. In this problem, the divisor is $x + 3$, so the opposite of its constant is -3. Separate it from the rest of the coefficients by writing it like this: $-3|$. Leave some space beneath everything and draw a horizontal line.

Critical Point

Even though synthetic division can only be applied when you're dividing by a linear binomial, it will be extremely useful later on in Chapter 14.

$$\underline{-3|\ \ 2 \quad 0 \quad -1 \quad 4}$$

The setup is complete, and it's time to get started. Take the leading coefficient of the dividend (2) and drop it below the horizontal line.

$$\begin{array}{r} -3|\ \ 2 \quad 0 \quad -1 \quad 4 \\ \hline 2 \end{array}$$

Multiply the number in the box (–3) by the number below the line (2) and write the result (–6) below the next coefficient (0).

$$
\begin{array}{r|rrrr}
-3 & 2 & 0 & -1 & 4 \\
& & -6 & & \\
\hline
& 2 & & &
\end{array}
$$

Combine the numbers in the second column (0 – 6 = –6) and write the result at the bottom of that column, below the horizontal line.

$$
\begin{array}{r|rrrr}
-3 & 2 & 0 & -1 & 4 \\
& & -6 & & \\
\hline
& 2 & -6 & &
\end{array}
$$

Repeat the process two more times, each time multiplying the number in the box by the new number below the line, writing the result in the next column, combining the numbers in that column, and writing the result below the horizontal line.

$$
\begin{array}{r|rrrr}
-3 & 2 & 0 & -1 & 4 \\
& & -6 & 18 & -51 \\
\hline
& 2 & -6 & 17 & -47
\end{array}
$$

The numbers below the line are the coefficients of the quotient starting with x^2, which is one less than the degree of the dividend: $2x^2 - 6x + 17$. The leftover number all the way on the right (–47) is the remainder, which is written just like it was in long division.

$$
2x^2 - 6x + 17 + \frac{-47}{x+3}
$$

A final answer of $2x^2 - 6x + 17 - \dfrac{47}{x+3}$ is equally correct—you're allowed to move the negative sign out of the numerator if you want to. You can check your answer the same way you did in long division. Multiply the quotient by the divisor and add the remainder; you should end up with the original dividend.

$$
\begin{aligned}
&\left(2x^2 - 6x + 17\right)(x+3) + (-47) \\
&= 2x^2(x) + 2x^2(3) + (-6x)(x) + (-6x)(3) + 17(x) + 17(3) - 47 \\
&= 2x^3 + 6x^2 - 6x^2 - 18x + 17x + 51 - 47 \\
&= 2x^3 - x + 4
\end{aligned}
$$

You've Got Problems
Problem 6: Calculate the quotient $(4x^3 - 2x^2 - 10x + 1) \div (x - 2)$.

The Least You Need to Know

- If two terms in a polynomial have variables that match exactly, they are like terms.

- Only like terms can be added or subtracted.

- Multiply polynomials by distributing each term of the left polynomial to each term of the right polynomial, one at a time.

- Long division correctly calculates any polynomial quotient, but synthetic division (which only works when the divisor is a linear binomial) is faster.

Chapter 11

Factoring Polynomials

In This Chapter

- Finding greatest common factors
- Recognizing factoring patterns
- Factoring trinomials using their coefficients
- Factoring difficult trinomials with the bomb technique

I remember the first day I tried to drive in reverse. I didn't think it would be any harder than driving forward (after all, walking backward is no harder than walking forward, so why should driving be any different?), but boy, was I wrong. Everything feels different when you're backing up. Turning the steering wheel left makes the front of the car go right, for one thing, and that takes a while to get used to. I still get a little antsy in reverse; it's a whole new ballgame.

Chapter 10 focused on multiplying polynomials (in my metaphor, the equivalent of driving), and now it's time to throw the whole process in reverse to learn factoring.

Why is factoring the reverse of multiplication, you ask? Factoring takes what was once a product and breaks it into pieces (called factors) that multiply together to get that product.

According to Chapter 10, $(x - 3)(x + 5) = x^2 + 2x - 15$. In this chapter, you'll start with $x^2 + 2x - 15$ and wind up with $(x - 3)(x + 5)$. At first, things will feel a little strange, like driving in reverse originally does, but you'll get used to it quickly. Just don't be afraid to knock a few mailboxes over in the process—it happens to everyone.

Greatest Common Factors

The *greatest common factor* (GCF) of a polynomial is the largest monomial that divides evenly into each term. It's very similar to the greatest common factor from Chapter 2 "Simplifying Fractions," except polynomial GCFs usually contain variables.

Here's how to calculate the GCF of a polynomial:

1. **Find the GCF of the polynomial's coefficients.** The GCF of the coefficients will be the coefficient of the GCF. Try and say that three times fast!

Talk the Talk

Factoring is the process of returning a polynomial product back into unmultiplied pieces called factors. A polynomial's **greatest common factor** is the largest monomial that divides evenly into each of the polynomial's terms.

2. **Identify common variable powers.** Look at the variables in each term of the polynomial. The GCF should contain the lowest power of every variable that appears in every term. That means each term will contain the variable raised to *at least* that exponent. That sounds tricky, but it's actually really easy, as you'll see in Example 1.

3. **Multiply.** The product of steps 1 and 2 above is the GCF of the polynomial.

Once you've found the GCF of a polynomial, you're ready for *factoring*. Write the GCF followed by a set of parentheses. Inside those parentheses, list what's left of each polynomial term after you divide it by the GCF. In other words, the parentheses show the polynomial with the GCF "sucked out."

Example 1: Factor the polynomial $6x^2y^3 - 12xy^2$.

Solution: Start by finding the GCF of the coefficients, 6 and 12. The largest number that divides evenly into both is 6. Therefore, 6 is the coefficient of the polynomial's GCF. Now it's time to figure out what variables appear in the GCF. Ask yourself, "What are the smallest powers of x and y?" The terms contain an x^2 and an x, of which x has the smaller power. Similarly, y^2 in the second term has a smaller power than y^3 in the first term.

Put it all together. The GCF has a coefficient of 6 and contains the variables xy^2, so $6xy^2$ is the GCF. Divide both terms by the GCF.

$$\frac{6x^2y^3}{6xy^2} + \frac{-12xy^2}{6xy^2}$$

$$= \frac{6}{6}x^{2-1}y^{3-2} - \frac{12}{6}x^{1-1}y^{2-2}$$

$$= xy - 2$$

You're almost done. Factor $6x^2y^3 - 12xy^2$ by rewriting it as the GCF multiplied by the terms you just got by dividing.

$$6xy^2(xy - 2)$$

It's easy to check your answer. Distribute $6xy^2$ to each term in the parentheses, and make sure you end up with the original expression, $6x^2y^3 - 12xy^2$.

$$6xy^2\left(xy - 2\right)$$

$$= 6xy^2\left(xy\right) + 6xy^2\left(-2\right)$$

$$= 6x^{1+1}y^{2+1} + \left(-12\right)xy^2$$

$$= 6x^2y^3 - 12xy^2$$

You've Got Problems

Problem 1: Factor the polynomial $9x^5y^2 + 3x^4y^3 - 6x^3y^7$.

Factoring by Grouping

Factoring by grouping is used only in very specific circumstances; and of all the factoring techniques in this chapter, you'll probably use it the least. However, when it's applicable, it gets the job done, and gets it done fast.

Factoring by grouping works best when you're given a polynomial with four terms that don't all share a greatest common factor. You'll split the large polynomial into two smaller pieces, each of which contains two terms with common factors. Then, you'll factor both of those smaller groups individually.

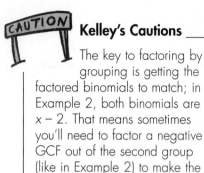

Kelley's Cautions

The key to factoring by grouping is getting the factored binomials to match; in Example 2, both binomials are $x - 2$. That means sometimes you'll need to factor a negative GCF out of the second group (like in Example 2) to make the binomials match.

Example 2: Factor the polynomial $2x^3 - 4x^2 - 3x + 6$.

Solution: Unfortunately, these terms don't have any factor in common (except 1, and that's true of any group of terms). Look at the polynomial as two groups, each containing two terms.

$$(2x^3 - 4x^2) + (-3x + 6)$$

Notice that the left group has a GCF of $2x^2$, so factor it out (like you did in Example 1).

$$= 2x^2(x - 2) + (-3x + 6)$$

You can factor 3 out of the second group of terms, but it's more useful to factor out -3. Why? When you do, the binomial in parentheses *exactly matches* the binomial you got when you factored the first group.

$$= 2x^2(x - 2) - 3(x - 2)$$

Now comes the slightly confusing part. Factor the binomial $(x - 2)$ out of both terms. Although factoring out something other than a monomial may feel a little weird, it's definitely allowed. Once you pull $(x - 2)$ out, all you're left with in the first term is $2x^2$; and in the second term, only -3 remains. Finish factoring by writing $(x - 2)$ followed by those leftovers in a second set of parentheses.

$$= (x - 2)(2x^2 - 3)$$

Check your work by multiplying the factors to make sure you get the original expression.

$$= x(2x^2) + x(-3) + (-2)(2x^2) + (-2)(-3)$$
$$= 2x^3 - 3x - 4x^2 + 6$$
$$= 2x^3 - 4x^2 - 3x + 6$$

You've Got Problems

Problem 2: Factor the polynomial $12x^4 + 6x^3 + 14x + 7$.

Special Factoring Patterns

In some cases, factoring a polynomial takes almost no effort at all. This appeals to me, because deep down, I am a very lazy man, and if I didn't have to feed myself or my family, I'd be happily working a meaningless job and living in squalor (as I proved beyond any reasonable doubt when I was a bachelor).

When it comes to factoring, you're allowed to be lazy and apply a formula if the polynomial follows one of three specific patterns. To spot these handy shortcut opportunities, keep your eyes peeled for perfect squares and perfect cubes.

A *perfect square* is what you get when you multiply something by itself. In other words, you create a perfect square by squaring something. For instance, $36w^4$ is a perfect square, because it's the result of something multiplied by itself: $(6w^2)(6w^2) = 36w^4$.

A *perfect cube* is created by multiplying something by itself two times (in other words, the result of cubing something). Therefore, $-8y^3$ is a perfect cube, because $(-2y)(-2y)(-2y) = -8y^3$.

Talk the Talk

A **perfect square** is created by multiplying something times itself, and a **perfect cube** is created by multiplying something times itself twice.

Here are the special factor patterns you should learn to recognize. Memorize the formulas, because trying to generate them from scratch is a waste of time.

- ◆ **The difference of perfect squares:** The expression $a^2 - b^2$ (one perfect square minus another) factors into $(a + b)(a - b)$. For example, the polynomial $x^2 - 16$ is a difference of perfect squares because $(x)^2 = x^2$ and $(4)^2 = 16$; both x^2 and 16 are the result of squaring something. Compare $x^2 - 16$ to the formula $a^2 - b^2$ to see that $x^2 = a^2$ and $16 = b^2$, so $x = a$ and $4 = b$. Plug those values of a and b into the difference of perfect squares pattern to factor.

$$a^2 - b^2 = (a+b)(a-b)$$
$$x^2 - 16 = (x+4)(x-4)$$

- ◆ **The difference of perfect cubes:** If two perfect cubes are subtracted, they can be factored as the product of a binomial and a trinomial: $a^3 - b^3 = (a - b)(a^2 + ab + b^2)$. For example, the polynomial $8x^3 - 27$ is a difference of perfect cubes; to apply the formula (and therefore, factor it), set $8x^3 - 27 = a^3 - b^3$, meaning $a = 2x$ and $b = 3$.

$$a^3 - b^3 = (a-b)(a^2 + ab + b^2)$$
$$(2x)^3 - (3)^3 = (2x-3)\left[(2x)^2 + (2x)(3) + (3)^2\right]$$
$$8x^3 - 27 = (2x-3)(4x^2 + 6x + 9)$$

◆ **The sum of perfect cubes:** The sum of two cubes factors almost exactly like the difference of two cubes—only a few signs are different: $(a^3 + b^3) = (a + b)(a^2 - ab + b^2)$. Consider the polynomial $y^3 + 64$, the sum of perfect cubes $a^3 + b^3$ when $a = y$ and $b = 4$.

$$a^3 + b^3 = (a+b)(a^2 - ab + b^2)$$

$$(y)^3 + (4)^3 = (y+4)\left[(y)^2 - (y)(4) + (4)^2\right]$$

$$y^3 + 64 = (y+4)(y^2 - 4y + 16)$$

 Kelley's Cautions

There is no factoring pattern for the *sum* of perfect squares, so you can't factor $a^2 + b^2$. Algebra students often factor it incorrectly as $(a + b)$ $(a + b)$. If you multiply $(a + b)(a + b)$, you get $a^2 + 2ab + b^2$, not $a^2 + b^2$.

Keep in mind that you can pull a greatest common factor out of a polynomial *before* you apply the three factoring patterns. In fact, you should always look for a greatest common factor (and if it exists, factor it out) before you even think about any other factoring technique. That way you're sure to get the fully factored version of the polynomial.

Example 3: Factor the polynomials.

(a) $16x^3 + 2y^3$

Solution: This feels like a sum of perfect cubes problem, thanks to the powers of 3, but it doesn't exactly fit the formula. Even though x^3 is a perfect cube, $16x^3$ isn't. (There's no rational number that equals 16 if multiplied by itself twice.) Same goes for $2y^3$.

Hope is not lost, however. The terms have a greatest common factor of 2, so factor it out.

$$2(8x^3 + y^3)$$

Kelley's Cautions

Always assume that you're expected to factor a polynomial completely. In other words, none of the final factors should be factorable.

Ignore that 2 outside the parentheses for now; the quantity inside fits the sum of perfect cubes formula if $a = 2x$ and $b = y$. Rewrite $8x^3 + y^3$ in factored form, leaving the GCF exactly where it is out front.

$$2(2x + y)(4x^2 - 2xy + y^2)$$

(b) $x^4 - 16$

Solution: This is a difference of perfect squares, and fits the formula $a^2 - b^2 = (a + b)(a - b)$ if $a = x^2$ and $b = 4$.

$$(x^2 + 4)(x^2 - 4)$$

If you were to end here, you'd technically get the problem wrong, because it's not factored *completely*. One of the factors, $x^2 - 4$, is also a perfect square and must be factored further: $x^2 - 4 = (x + 2)(x - 2)$.

$$(x^2 + 4)(x + 2)(x - 2)$$

You've Got Problems

Problem 3: Factor the polynomial $5x^2 - 125$.

Factoring Trinomials Using Their Coefficients

Of all the expressions you'll factor, quadratic trinomials (polynomials with three terms and a highest exponent of 2) are probably the most common, so the rest of this chapter will focus on them. This section deals with trinomials that have a leading coefficient of 1—the final section of the chapter, "Factoring with the Bomb Method," takes on polynomials with different leading coefficients.

A quadratic trinomial with a leading coefficient of 1 looks like $x^2 + ax + b$, where a and b are integers. Your goal is to factor that trinomial into two binomials that look like this:

$$\left(x + \boxed{?}\right)\left(x + \boxed{?}\right)$$

Basically, all you have to do is figure out what numbers go in those $\boxed{?}$ boxes. Here's the trick: If the trinomial $x^2 + ax + b$ is factorable, then two numbers will exist whose sum is a (the coefficient of the x-term) and whose product is b (the constant). Once you know what a and b are, plug each one into its own box (it doesn't matter which one), and you've factored the trinomial.

Critical Point

When you're trying to figure out what goes into the $\boxed{?}$ boxes to factor $x^2 + ax + b$, pay special attention to the signs of a and b. If b is positive, then the mystery numbers are either both positive or both negative. If b is negative, then one of the mystery numbers must be positive and the other must be negative.

Example 4: Factor the polynomials.

(a) $x^2 + 7x + 12$

> **Solution:** Ask yourself, "What two numbers add up to 7 and multiply to give you 12?" Because 12 is positive, the mystery numbers are either both positive or both negative. However, the sum of the mystery numbers is positive, so the

numbers are positive as well. (There's no way to add two negative numbers and get a positive sum.)

If the answer doesn't dawn on you right away, set one mystery number equal to 1. In this problem, the numbers have to add up to 7, so that means the other mystery number is 6. However, $6 \cdot 1 \neq 12$, so change the numbers from 1 and 6 to 2 and 5; those numbers also have a sum of 7. Unfortunately, $5 \cdot 2 = 10$, which doesn't equal 12 either, but at least it's closer to 12 than $1 \cdot 6$ was.

Continue to increase your guess for the smaller mystery by 1. In other words, now it's time to try 3 and 4. This looks promising: $3 + 4 = 7$ and $4 \cdot 3 = 12$. Eureka! Plug those numbers into $\left(x + \boxed{?}\right)\left(x + \boxed{?}\right)$ and you're done: $(x + 3)(x + 4)$. Multiplication is commutative, so the answer $(x + 4)(x + 3)$ is equally correct.

(b) $w^2 - 3w - 54$

Solution: Even though this has w's instead of x's, the process is the same. The constant -54 is negative, so that means one of the mystery numbers is positive and the other is negative. Because they add up to a negative number (-3), the negative mystery number is bigger than the positive number.

Critical Point

Another way to find the mystery numbers in Example 4(b) is to start with the constant, rather than the x-coefficient. Instead of trying pairs of numbers that add up to -3, try pairs of numbers whose product is -54.

If you can't figure out where to start, use 1 as a mystery number—like in Example 4(a)—and keep increasing it by 1 until you hit pay dirt: $-4 + 1$, $-5 + 2$, $-6 + 3$, etc. You'll eventually hit the numbers -9 and 6, which have the correct sum and product ($-9 + 6 = -3$ and $-9(6) = -54$), so the factored form of the polynomial is $(x - 9)(x + 6)$.

You've Got Problems

Problem 4: Factor the polynomial $2x^3 - 24x^2 + 64x$.

Factoring with the Bomb Method

When the leading coefficient of a quadratic trinomial isn't 1, it's time to go to the heavy artillery and break out the bomb technique to factor. To be honest, I am the only one who uses this strangely explosive name to describe the process—I learned the bomb method many moons ago under the uninspiring name "factoring by decomposition" and decided to rename it.

Not many algebra teachers even know this technique exists, and instead advise you to "play around" with numbers until you come up with a pair of binomials that works. That is very bad advice, especially when faced with tough factoring problems. After all, just about everything else you learn in algebra has very specific and rigid steps you have to follow in *exactly* the right order, or else the ghosts of long-dead mathematicians rise from their graves, kick you in the shins, and mark points off your test.

Critical Point

Unfactorable polynomials, like unfactorable numbers, are described as prime.

Here are the steps that light the fuse on the bomb technique to factor the polynomial $ax^2 + bx + c$:

1. **Factor out the GCF, if there is one.** This should always be your first step in a factoring problem.

2. **Find two mystery numbers.** These are similar to the numbers you sought when factoring simpler trinomials. You still want the sum to be the x-coefficient (b), but now you want the product to equal $a \cdot c$, the leading coefficient times the constant.

3. **Replace the x-coefficient.** Rewrite the polynomial, but where b once stood, write the sum of the two mystery numbers in parentheses. This powerful little clump of parentheses will explode, splitting the trinomial up into two binomials—that's why I call this the bomb method.

4. **Distribute x through the parentheses.** There's still an x next to the parentheses. Multiply each of the mystery numbers by x and ditch the parentheses.

5. **Factor by grouping.** As described earlier in the chapter, factor by grouping to change the four-term polynomial into a product of linear binomials.

The bomb method takes practice. Make up your own trinomials to factor by multiplying simple linear binomials together; and then try to factor the product back into those binomials.

Example 5: Factor the polynomial $6x^2 - x - 12$.

Solution: This polynomial has no GCF, so skip right to calculating the mystery numbers. They should add up to -1 and have a product of $6(-12) = -72$. The only numbers to meet those requirements are -9 and 8. Replace the x-coefficient -1 by adding those numbers in parentheses.

$$6x^2 + (-9 + 8)x - 12$$

Distribute the x to the right of the parentheses.

$$6x^2 - 9x + 8x - 12$$

Now you can factor by grouping to finish.

$$\left(6x^2 - 9x\right) + \left(8x - 12\right)$$
$$= 3x\left(2x - 3\right) + 4\left(2x - 3\right)$$
$$= \left(2x - 3\right)\left(3x + 4\right)$$

You've Got Problems

Problem 5: Factor the polynomial $4x^2 + 23x - 6$.

The Least You Need to Know

- Start every factoring problem by checking for a greatest common factor.

- Differences of perfect squares can be factored, as can sums and differences of perfect cubes. However, there is no pattern that factors sums of perfect squares.

- When factoring a quadratic trinomial with a leading coefficient of 1, find two numbers that add up to the x-coefficient of the trinomial and multiply to give you the constant.

- In the bomb method for factoring trinomials of type $ax^2 + bx + c$, you replace b with the sum of two mystery numbers that equal b when added and equal $a \cdot c$ when multiplied.

Chapter 12

Wrestling with Radicals

In This Chapter

♦ Simplifying radical expressions

♦ Adding, subtracting, multiplying, and dividing radical expressions

♦ Solving simple equations that contain radicals

♦ Dealing with imaginary numbers

Exponents are old news. You've squared things, cubed things, occasionally raised things to a negative power, and even gotten into a shoving match with a variable raised to the fifth power once, outside a shady nightclub on the edge of town. (If his coefficient hadn't been there, who knows what might have happened?)

Let's be frank. Exponents have been strutting around and showing off. They're at the top of the food chain, with no natural enemies. Sure, addition's great, but it has to look out for its arch nemesis subtraction, and multiplication has never been the same since division showed up and started talking to his girlfriend. Finally, the worm has turned, and exponents are about to be put in their respective places; this chapter deals with radicals, little symbols that keep exponential powers in check that (coincidentally) look like little check marks.

Introducing the Radical Sign

A *radical expression* looks like $\sqrt[a]{b}$ (read "the ath root of b"), and consists of three parts:

◆ *Radical symbol:* the symbol that looks like a check mark with an elongated horizontal line stuck to the end of it.

Talk the Talk

The **radical expression** $\sqrt[a]{b}$ has three major features: the **radical symbol** (which looks like a check mark), the **index** (the small number above the sharp point in the radical symbol), and the **radicand,** which is beneath the horizontal bar of the radical

◆ *Index:* the small number tucked inside the check mark portion of the radical sign; in the expression $\sqrt[a]{b}$, a is the index.

◆ *Radicand:* the quantity written inside the radical symbol, beneath its horizontal roof; b is the radicand of the radical expression $\sqrt[a]{b}$.

Most of the radicals you'll deal with will have an index of 2, and are called square roots. When no explicit index is written in a radical expression (such as $\sqrt{13}$), you automatically assume that the index is 2.

Simplifying Radical Expressions

Think of a radical symbol like a prison and the pieces of the radicand as inmates. Not all the prisoners are doomed to a life sentence, trapped inside the dank (and foul-smelling) radical big house—there is a chance of parole. However, in order to be released from the radical sign, you must meet its parole requirements.

Talk the Talk

Radicals with an index of 2 are called **square roots,** although an index of 2 is rarely written explicitly ($\sqrt{5x}$ not $\sqrt[2]{5x}$). Radicals with an index of 3 are called cube roots. While an index can be any natural number, only these two kinds of radicals have special names.

Specifically, a radical will only release things raised to a power that matches its index. So *square roots* will only release pieces of its radicand that are raised to the second power, and a radical with an index of 5 will release only things raised to the fifth power. When you simplify radicals, you parole the factors within that meet the requirements, and leave the rest inside to rot.

Example 1: Simplify the radical expressions.

(a) $\sqrt[3]{16x^4 y^6}$

Solution: Start by factoring the coefficient—write it as a product of smaller numbers. Because the index of the radical is 3, you want the factors to be powers of 3 whenever possible. Therefore, instead of writing 16 as $16 \cdot 1$ or $4 \cdot 4$, write it as $8 \cdot 2$, or $2^3 \cdot 2$, where one of the factors is a perfect cube $(8 = 2^3)$.

Critical Point

To make simplifying radicals easier, memorize the first 15 perfect squares (1, 4, 9, 16, 25, 36, 49, 64, 81, 100, 121, 144, 169, 196, 225) and the first 5 perfect cubes (1, 8, 27, 64, 125).

$$\sqrt[3]{2^3 \cdot 2 \cdot x^4 y^6}$$

Now turn your attention to the variables. You can rewrite x^4 as $x^3 \cdot x$ (because $x^3 \cdot x = x^{3+1} = x^4$), so it contains an exponent of 3. Luckily, y^6 is a perfect cube $(y^2 \cdot y^2 \cdot y^2 = y^6)$, so write it using that all-important power of 3 also: $\left(y^2\right)^3$.

Critical Point

Why do you rewrite y^6 as $\left(y^2\right)^3$ in Example 1(a)? You're trying to make groups of three things, so that they can be released from the radical. It might help to think of $\left(y^2\right)^3$ as a group of three y^2's multiplied together, and $\left(y^2\right)^3 = y^6$ thanks to the exponential rule from Chapter 3 that says $\left(x^a\right)^b = x^{ab}$.

$$\sqrt[3]{2^3 \cdot 2 \cdot x^3 \cdot x \cdot \left(y^2\right)^3}$$

Of all the factors in the radicand, only 2^3, x^3, and $\left(y^2\right)^3$ are raised to the third power. Yank them out in front of the radical, stripping away the third power as they exit the prison, leaving the factors 2 and x inside.

$$\left(2xy^2\right)\sqrt[3]{2x}$$

(b) $\sqrt{18x^2 y^3}$

Solution: There's no index written on this radical, which means that it's a square root and the index is understood to be 2. That means you want as much of the radicand as possible raised to the second power. The coefficient 18 has only one factor that's a perfect square (9), so rewrite 18 as the product $2 \cdot 9$ (or $2 \cdot 3^2$). The

x^2-term already has an exponent of 2, but you should rewrite the y^3 term as $y^2 \cdot y$, to identify y^2 as a candidate for release.

$$\sqrt{2 \cdot 3^2 \cdot x^2 \cdot y^2 \cdot y}$$

Pull everything with an exponent of 2 outside the radical (and take away the power as you do), leaving what's left as the new radicand.

$$3|xy|\sqrt{2y}$$

You've got to be wondering where the heck those absolute value signs came from. I kind of sprung them on you, and I apologize. There's a rule in algebra that says if you ever have the expression $\sqrt[n]{x^n}$ (the power of the exponent matches the index of the radicand), and n is even, then the simplified result is $|x|$. Here's why: you always want an even-powered root to have a positive answer, and those absolute values make sure that no matter what the variable equals, the answer will be positive.

Kelley's Cautions

If you have a variable in a radicand that's raised to the same even power as the index, you should surround the paroled variable with absolute value symbols.

You're releasing both x^2 and y^2 from this square root, and those exponents match the index of the radical, so you've got to toss them inside absolute value signs once they're paroled.

You've Got Problems

Problem 1: Simplify the radical expression $\sqrt{300x^6y^3}$.

Unleashing Rational Powers

Radicals can actually be written without a radical sign by means of fractional exponents. The expression $a^{1/2}$ (read "a to the one-half power") is equivalent to $\sqrt{a}$, and $b^{1/3}$ means the same thing as $\sqrt[3]{b}$. The fractional exponent doesn't always have a numerator of 1 though—you can use other numbers to create more complicated radical expressions.

Here's how rational exponents work: the expression $x^{a/b}$ is equal to both $\sqrt[b]{x^a}$ and $\left(\sqrt[b]{x}\right)^a$. Even though they look different, both mean the exact same thing, so it doesn't matter which one you choose. However, you'll usually find that the first one is more useful for simplifying variables and the second is much handier for numbers.

Example 2: Simplify the expressions.

(a) $64^{2/3}$

Solution: Rewrite $64^{2/3}$ as $\left(\sqrt[3]{64}\right)^2$. The order of operations tells you to simplify parentheses first, and because the radical has an index of 3, you're looking for perfect cubes. Luckily, $64 = 4^3$: $\left(\sqrt[3]{64}\right)^2 = \left(\sqrt[3]{4^3}\right)^2$. When 4 is raised to the third power inside a radical of index 3, it gets paroled: $\left(\sqrt[3]{4^3}\right)^2 = 4^2 = 16$. Therefore, $64^{2/3} = 16$.

If you had rewritten $64^{2/3}$ as $\sqrt[3]{64^2} = \sqrt[3]{4,096}$, the answer would still have been 16. It's just harder to identify that 4096 as a perfect cube ($16^3 = 4096$).

(b) $4^{5/2}$

Solution: Rewrite the expression as $\left(\sqrt{4}\right)^5$. Happily, 4 is a perfect square: $2^2 = 4$.

$$\left(\sqrt{4}\right)^5 = \left(\sqrt{2^2}\right)^5 = 2^5 = 32$$

Critical Point

The denominator of a fractional power is the index of the equivalent radical, and the numerator is a power, either of the radicand or of the entire radical.

You've Got Problems

Problem 2: Simplify the expression $25^{3/2}$.

Radical Operations

The phrase "radical operations" doesn't mean "extreme medical procedures," like having your arm removed and replaced with an otter, or perhaps having your cousin Irving surgically grafted to your left side, so that you become the world's first man-made conjoined twins. Instead, it refers to the much more boring concepts of adding, subtracting, multiplying, and dividing radicals.

There are specific rules you have to follow when radical expressions are involved, just like there are special rules governing polynomial expressions. Once again, you'll find that multiplying and dividing radicals are much easier than adding and subtracting them.

Addition and Subtraction

Chapter 10 insisted that you could add or subtract only like terms—terms with the exact same variables. Radicals operate in a very similar way. In order to add or subtract them, they must be *like radicals*, containing the exact same radicand and index.

If you are supposed to add or subtract radicals that contain different radicands, don't panic. Try to simplify the radicals—that usually does the trick.

Example 3: Simplify the expression $3\sqrt{2xy} - 2\sqrt{50xy}$.

Solution: Notice that the second radical can be simplified, because $50 = 25 \cdot 2$ and 25 is a perfect square.

$$2\sqrt{50xy} = 2\sqrt{25 \cdot 2 \cdot xy} = 2\sqrt{5^2 \cdot 2 \cdot xy} = 2 \cdot 5\sqrt{2xy} = 10\sqrt{2xy}$$

The original problem now looks like this:

$$3\sqrt{2xy} - 10\sqrt{2xy}$$

Talk the Talk

Like radicals have matching radicands and indices, like $6\sqrt[5]{2x^2 y}$ and $-9\sqrt[5]{2x^2 y}$.

Without even trying, you've created a pair of like radicals—the radicands and indices match. Because they're like radicals, you can combine their coefficients ($3 - 10 = -7$) and tack on the common radical expression $\sqrt{2xy}$. The final answer is $-7\sqrt{2xy}$.

You've Got Problems

Problem 3: Simplify the expression $\sqrt[3]{8x^4} + 4x\sqrt[3]{x}$.

Multiplication

If two radicals with the same index are multiplied together, the result is just the product of the radicands beneath a single radical with a matching index. Translation: if you're multiplying radicals with matching indices, just multiply the things inside the radical signs, and write the result under a radical sign with the same index the original radicals had.

$$\left(\sqrt{x}\right)\left(\sqrt{y}\right) = \sqrt{xy} \qquad \left(\sqrt[3]{7}\right)\left(\sqrt[3]{w^2}\right) = \sqrt[3]{7w^2}$$

Example 4: Simplify the expression $\left(\sqrt[3]{9x^4 y^5}\right)^2$

Solution: Squaring a radical means multiplying it by itself.

$$\left(\sqrt[3]{9x^4y^5}\right)\left(\sqrt[3]{9x^4y^5}\right)$$

Multiply the radicands together and write the product beneath a radical sign with the same index (3).

$$\sqrt[3]{9 \cdot 9 \cdot x^4 \cdot x^4 \cdot y^5 \cdot y^5} = \sqrt[3]{81x^8y^{10}}$$

Simplify the radical.

$$\sqrt[3]{3^3 \cdot 3 \cdot \left(x^2\right)^3 \cdot x^2 \cdot \left(y^3\right)^3 \cdot y} = 3x^2y^3\sqrt[3]{3x^2y}$$

You've Got Problems

Problem 4: Simplify the product $\left(\sqrt{12x^2y}\right)\left(\sqrt{3xy}\right)$.

Division

The quotient of two radicals with the same index can be rewritten beneath a single radical sign, just like the product of those radicals. In other words, the expression

$\sqrt{x} \div \sqrt{y}$ is equivalent to $\sqrt{\dfrac{x}{y}}$. However, there is a new concern that surfaces when

you're dealing with radical division—the presence of a radical in the denominator of your final answer.

For a long time, it has been considered bad etiquette to leave a radical in the denominator of a final answer. After all, irrational numbers are long and ugly decimals. It's not as though you're actually expected to divide that horrible decimal into the numerator or anything, but even the *prospect* of such a gross division problem has culminated in decades of peer pressure to eliminate any denominator radicals in a process called

rationalizing the denominator. It's a final, and easy, step some teachers require and others (like me) don't. Make sure to ask your instructor whether you are expected to rationalize the denominators in your solutions.

 Talk the Talk

The process of removing radicals from the denominator is called **rationalizing the denominator.**

Example 5: Simplify the expression $\sqrt{12x^7 y} \div \sqrt{8xy^3}$ and rationalize the quotient.

Solution: Write the quotient as a fraction beneath a single radical sign.

$$\sqrt{\frac{12x^7 y}{8xy^3}}$$

Simplify the fraction.

Critical Point

You're allowed to multiply a fraction by anything divided by itself, because that's technically the same thing as multiplying by 1. (Anything divided by itself equals 1.)

$$\sqrt{\frac{12}{8} x^{7-1} y^{1-3}} = \sqrt{\frac{3}{2} x^6 y^{-2}} = \sqrt{\frac{3x^6}{2y^2}}$$

Write the numerator and denominator as separate radicals and simplify them.

$$\frac{\sqrt{3x^6}}{\sqrt{2y^2}} = \frac{\sqrt{3(x^3)^2}}{\sqrt{2(y)^2}} = \frac{x^3 \sqrt{3}}{|y|\sqrt{2}}$$

The denominator of the fraction contains the radical $\sqrt{2}$; to eliminate it, multiply both the numerator and the denominator by $\sqrt{2}$.

$$\frac{x^3 \sqrt{3}}{|y|\sqrt{2}}\left(\frac{\sqrt{2}}{\sqrt{2}}\right) = \frac{x^3 \sqrt{6}}{|y|\sqrt{4}}$$

The radical in the denominator now contains a perfect square—simplify it to rationalize the denominator.

$$\frac{x^3 \sqrt{6}}{|y|\sqrt{4}} = \frac{x^3 \sqrt{6}}{2|y|}$$

You've Got Problems

Problem 5: Simplify the expression $\sqrt{2x^2y^3} \div \sqrt{18x^3y^2}$ and rationalize the quotient.

Solving Radical Equations

Have you ever read *Alice in Wonderland?* It had a (slightly less popular) sequel called *Through the Looking Glass*, in which Alice passes through a mirror into a world that's a weird reflection of her own. Some physicists have likened the relationship between these worlds to the relationship between matter and antimatter.

Without getting too nerdy or discussing *Star Trek* propulsion systems (which are theoretically powered by matter/antimatter engines—too late, I discussed them!), I'll suffice it to say that you should never have matter and antimatter at the same social gathering, because if they come in contact with one another, there'll be a cataclysmic explosion. They are the exact antitheses of one another, so much so that when they touch, they cancel one another out in the most permanent way possible: detonation. A slightly less violent, but similar, relationship exists between me and the girl I dated in high school, but that's neither here nor there.

The same relationship exists between radicals with index n and exponential expressions with power n. If they meet, they'll say, "It wasn't me that changed, it was you" (actually, that sounds more like my ex-girlfriend) and promptly explode, leaving behind only the contents of the radicand, a pile of smoke, and some embarrassing old love letters you can't remember writing.

This explosive property can be used to solve equations containing radicals, or even to power the concert pyrotechnics of heavy metal bands like Limozeen and Taranchula. In the interest of space, I'll focus on the first application.

Example 6: Solve the equation $\sqrt[3]{2x-1}+3=6$.

Solution: The variable x is trapped inside a radical sign; you'll need to free it with some specially designed explosives. Before you set the charge, get everything but the radical out of harm's way—isolate the radical by subtracting 3 from both sides of the equation.

$$\sqrt[3]{2x-1}=3$$

Talk the Talk

The heavy metal band names are a tribute to Strong Bad, a character that lives at the website www.homestar runner.com. If you haven't happened upon this website yet, make sure you pay them a visit, and tell Marzipan I said hi.

To destroy this radical with index 3, raise it to the third power. (If it had an index of 5, you'd raise both sides to the fifth power—match the exponent to the index of the doomed radical.) To keep the equation balanced, you should raise the right side to the third power as well.

$$\left(\sqrt[3]{2x-1}\right)^3 = 3^3$$

All that remains on the left side of the equation are the smoldering contents of the radicand. The resulting equation is very simple to solve.

$$2x-1=27$$
$$2x=28$$
$$x=14$$

You can check your answer by substituting 14 for x in the original problem.

$$\sqrt[3]{2x-1}+3=6$$
$$\sqrt[3]{2(14)-1}=3$$
$$\sqrt[3]{27}=3$$
$$3=3$$

The statement you end up with (3 = 3) is true, which means you got the right answer.

You've Got Problems

Problem 6: Solve the equation $2\sqrt{x-3}=8$.

When Things Get Complex

So far in this chapter, the vast majority of the radicands have been positive. They can't help it—they're just upbeat, and there's nothing wrong with that. However, you do need to know how to handle it when a bit of negativity creeps in.

Sometimes a negative is no problem. For example, when the index of a radical is odd, a negative radicand is completely valid; $\sqrt[3]{-8x^3}$ simplifies to $-2x$ because $(-2x)(-2x)(-2x) = -8x^3$. A negative thing multiplied by itself an odd number of times will also be negative. However, if a negative lurks inside a radical with an even index, there's trouble.

While $\sqrt{16}$ is easy to simplify $\left(\sqrt{16}=\sqrt{4^2}=4\right)$, the expression $\sqrt{-16}$ is not. You can't square a number and end up with something negative. A product of two numbers is negative only when the numbers have different signs, and there's no way a number can have a different sign than itself!

There's Something in Your *i*

One great thing about mathematicians is that they're good at inventing stuff that makes impossible things possible. One of their handiest inventions is *i*, a letter that's the solution to a problematic negative radicand. The letter *i* is short for "imaginary number," and has the value $i = \sqrt{-1}$.

A number containing *i* (such as $2i$ or $-5i$) is called an *imaginary number*, and numbers of the form $a + bi$ (where *a* and *b* are real numbers) is a *complex number*. Basically, a complex number is made up of an imaginary part, bi, added to or subtracted from a real part, a. For example, in the complex number $4 - 7i$, the real part is 4 and the imaginary part is $-7i$.

Every complex number $a + bi$ has a *conjugate* equal to $a - bi$. The only difference between a complex number and its conjugate is the sign of the imaginary part. That means the conjugate of $4 - 7i$ would be $4 + 7i$.

Talk the Talk

An **imaginary number**, bi, is the product of a real number b and the imaginary value $i = \sqrt{-1}$. A **complex number** has the form $a + bi$, where a and b are real numbers. Every complex number is paired with a **conjugate**, $a - bi$, which matches the complex number exactly, except for the sign of its imaginary part bi.

Here's something to chew on: If $i = \sqrt{-1}$, then $i^2 = \left(\sqrt{-1}\right)^2 = -1$. (If a radical is raised to an exponent that matches its index, both disappear, leaving behind only what was beneath the radical—in this case -1.) It's extremely weird that squaring produces a negative number, but that's the quirky and endearing role imaginary numbers play.

Example 7: Simplify the expressions.

(a) $\sqrt{-40}$

    Solution: Rewrite $\sqrt{-40}$ as $\sqrt{-1 \cdot 4 \cdot 10}$. Remove the perfect square ($4 = 2^2$) from the radical, and rewrite $\sqrt{-1}$ as i: $\sqrt{-1 \cdot 2^2 \cdot 10} = 2i\sqrt{10}$.

(b) i^5

    Solution: Rewrite i^5 using as many i^2 factors as possible.

$$i^5 = i^{2+2+1} = i^2 \cdot i^2 \cdot i$$

    Recall that $i^2 = -1$.

$$i^2 \cdot i^2 \cdot i = (-1)(-1)i = i$$

    Therefore, $i^5 = i$.

Simplifying Complex Expressions

Adding, subtracting, multiplying, and dividing complex numbers are similar to the methods used in Chapter 10 for polynomials, except when it comes to division. Here's a quick rundown describing how the four major operations work with complex numbers:

♦ **Addition:** Imaginary numbers contain the same variable, i, so treat them as like terms; add the real parts and the imaginary parts separately.

$$(3 - 4i) + (2 + 9i) = (3 + 2) + (-4i + 9i) = 5 + 5i$$

♦ **Subtraction:** Once you distribute the negative sign, all that's left is a simple addition problem.

$$(3 - 4i) - (2 + 9i) = (3 - 4i) + (-2 - 9i) = (3 - 2) + (-4i - 9i) = 1 - 13i$$

♦ **Multiplication:** Just like in any product of binomials, you distribute each term in the first complex number to each term in the second complex number.

$$(3 - 4i)(2 + 9i) = 3(2) + 3(9i) + (-4i)(2) + (-4i)(9i)$$
$$= 6 + 27i - 8i - 36i^2$$

Replace i^2 with -1 and combine like terms.

$$= 6 + (27i - 8i) - 36(-1)$$
$$= 42 + 19i$$

♦ **Division:** Good news! You don't have to perform long or synthetic division to calculate the quotient of complex numbers. To divide $1 - i$ by $2 + 7i$, start by writing the quotient as a fraction.

$$\frac{1 - i}{2 + 7i}$$

Multiply the numerator and the denominator by the conjugate of the denominator.

$$\frac{1 - i}{2 + 7i}\left(\frac{2 - 7i}{2 - 7i}\right) = \frac{(1 - i)(2 - 7i)}{(2 + 7i)(2 - 7i)}$$

You've created factors in the denominator that, when multiplied, turn into a difference of perfect squares. All of the imaginary terms will be eliminated from the denominator, which is why you multiplied by the conjugate in the first place.

Critical Point _____

No simplified complex number will ever contain i^2. Always replace i^2 with -1.

$$\frac{(1-i)(2-7i)}{(2+7i)(2-7i)} = \frac{2-7i-2i+7i^2}{4-14i+14i-49i^2} = \frac{2-9i+7(-1)}{4-49(-1)} = \frac{-5-9i}{53}$$

Complex numbers are usually written in the form $a + bi$, so divide both terms of the numerator by the denominator.

$$= -\frac{5}{53} - \frac{9i}{53}$$

You've Got Problems

Problem 8: Given the complex numbers $c = 3 - 4i$ and $d = 8 + i$, calculate the following:

(a) $c + d$

(b) $c - d$

(c) $c \cdot d$

(d) $c \div d$

The Least You Need to Know

♦ To simplify radicals, look for factors inside the radicand with exponents that match the index.

♦ You can add or subtract only like radicals.

♦ The expression $x^{a/b}$ can be rewritten as either $\sqrt[b]{x^a}$ or $\left(\sqrt[b]{x}\right)^a$.

♦ The complex number $a + bi$ is the sum of a real number a and an imaginary number bi.

♦ The conjugate of the complex number $a + bi$ is $a - bi$.

Chapter

13

Quadratic Equations and Inequalities

In This Chapter

- Finding solutions by factoring
- Completing the square
- Applying the quadratic formula
- Solving and graphing simple quadratic inequalities

When I taught math at a public high school, I also coached track and field. It may shock you that a math geek was doling out athletic advice, and quite honestly it should. I have never and will never be known for my nimble, dexterous reflexes. In fact, I actually got kicked out of a gymnastics program when I was seven, because they thought I was going to *paralyze* myself. As it turns out, I was the exact opposite of a cat. Whereas most cats can right themselves during a fall and land on their feet, I always managed to land on my head.

Regardless, the school administration saw fit to make me coach of the track team, and immediately assigned me the task of training the hurdlers. (I tried to hurdle only once as a coach, and I think I must have knocked myself out doing it, because when I regained consciousness, my face was planted firmly in the track. Interestingly enough, my feet were still propped up on the hurdle.) Even though I was not the most proficient coach in the world, I did learn one thing—always start training runners with the hurdles at their lowest possible setting. There's no sense starting a new athlete with the hurdle at its actual, intimidating race height right away. You start easy, and then slowly build the difficulty along the way.

Earlier chapters covered just about everything you ever wanted to know about linear equations (and a bunch of stuff I'm sure you could have lived a very long and happy life without knowing), so it's time to move the hurdles up a notch and throw in some polynomials of degree two. Solving linear and quadratic equations requires completely different techniques, but the good news is that you've got three different methods to choose from. By the end of the chapter, you'll leap like a gazelle over the new, higher hurdle, and I'll be quite proud of you, even though I'm down here, unconscious on the ground.

Solving Quadratics by Factoring

If you can transform an equation into a factorable quadratic polynomial, it is very simple to solve. Even though this technique will not work for all quadratic equations, when it does, it is by far the quickest and simplest way to get an answer. Therefore, unless a problem specifically tells you to use another technique, you should try this one first. If you get a prime (unfactorable) polynomial, you can always shift to one of the other techniques, either completing the square or the quadratic formula, covered later in this chapter.

To solve a quadratic equation by factoring, follow these steps:

1. **Set the equation equal to 0.** Move *all* of the terms to the left side of the equation by adding or subtracting them, as appropriate, leaving only 0 on the right side of the equation.

2. **Factor the polynomial completely.** Use the techniques from Chapter 11; remember to factor out the greatest common factor first.

3. **Set each of the factors equal to 0.** This will create a couple of little equations whose left sides are the factors and whose right sides are each 0. You should separate these equations with the word "or."

How'd They Do That?

If you have the equation $(x - a)(x - b) = 0$, Step 3 tells you to change that into the twin equations:

$$x - a = 0 \quad \text{or} \quad x - b = 0$$

Are you wondering why that's allowed? It's thanks to something called the *zero product property*. Think about it this way: if two things are multiplied together—in this case the quantities $(x - a)$ and $(x - b)$—and the result is 0, then at least one of those quantities must equal 0! There's no way to multiply two or more things and get 0 unless at least one of those things actually equals 0.

---

4. **Solve the small equations and check your answers.** Each of the solutions to the tiny, little equations is also a solution to the original equation. However, to make sure they actually work, you should plug them back into that original equation and verify that you get true statements.

The hardest part of this technique is the factoring itself, and since that's not a new concept, solving quadratic equations this way is very simple and straightforward.

Example 1: Solve the equations, and give all possible solutions.

(a) $x^2 - 6x + 9 = 0$

Solution: This equation is already set equal to 0, so start by factoring the left side.

$$(x - 3)(x - 3) = 0$$

Now set each factor equal to 0.

$$x - 3 = 0 \qquad x - 3 = 0$$
$$\text{or}$$
$$x = 3 \qquad x = 3$$

Because both factors were the same, the solutions were the same as well. Therefore, the equation $x^2 - 6x + 9 = 0$ only has one valid solution: $x = 3$. A solution that appears twice is called a *double root*.

Talk the Talk

A **double root** is a repeated solution for a polynomial equation, the result of a repeated factor in the polynomial.

Check to make sure that 3 is a valid answer by plugging it back into the original equation.

$$x^2 - 6x + 9 = 0$$
$$3^2 - 6(3) + 9 = 0$$
$$9 - 18 + 9 = 0$$
$$0 = 0$$

Because 0 = 0 is a true statement, the answer is correct.

(b) $3x^2 + 10x = -4x + 24$

Solution: Set the equation equal to 0 by adding $4x$ to, and subtracting 24 from, both sides.

$$3x^2 + (10x + 4x) + (-24) = (-4x + 4x) + (24 - 24)$$
$$3x^2 + 14x - 24 = 0$$

Factor the trinomial using the bomb method from Chapter 11. The two mystery numbers you're looking for are -4 and 18.

$$3x^2 + (-4 + 18)x - 24 = 0$$
$$3x^2 - 4x + 18x - 24 = 0$$
$$x(3x - 4) + 6(3x - 4) = 0$$
$$(3x - 4)(x + 6) = 0$$

Set each factor equal to 0 and solve.

$$3x - 4 = 0 \qquad \qquad x + 6 = 0$$
$$x = \frac{4}{3} \qquad \text{or} \qquad x = -6$$

Substitute the solutions into $3x^2 + 14x - 24 = 0$ to ensure they are valid.

$$3\left(\frac{4}{3}\right)^2 + 14\left(\frac{4}{3}\right) - 24 = 0$$
$$3\left(\frac{16}{9}\right) + \frac{56}{3} - 24 = 0$$
$$\frac{16}{3} + \frac{56}{3} - 24 = 0$$
$$\frac{72}{3} - 24 = 0$$
$$24 - 24 = 0$$

$$3(-6)^2 + 14(-6) - 24 = 0$$
$$3(36) - 84 - 24 = 0$$
$$108 - 108 = 0$$

You've Got Problems
Problem 1: Identify the solutions to the equation $4x^3 = 25x$.

Completing the Square

Solving a quadratic equation by *completing the square* forces a perfect square into the equation, which can be easily eliminated using a square root. It may not be the easiest way to solve the equation, but its strength lies in its predictability. The process of completing the square is like that guy you knew in high school whose parents made him take his cousin to the prom. Sure it was awkward, and neither of them had much fun, but at least their parents knew exactly how the evening was going to end. (After all, this was his *cousin*, so there was no prospect of post-prom "extracurricular activities.")

Here's the good news: completing the square will *always* work, unlike the factoring method, which is useless when the quadratic is prime. However, you need to master one skill before you can complete the square: using roots to eliminate exponents in equations.

Talk the Talk

This procedure is called **completing the square** because, by adding 16 to both sides in Example 2, you're creating a trinomial that's a perfect square. It factors into $(x + a)^2$, where a is the number from the bug's midsection.

Solving Basic Exponential Equations

According to Chapter 12, raising both sides of an equation to the nth power eliminates a radical with index n. In other words, to solve the equation $\sqrt[4]{x-1} = 3$, you should raise both sides to the fourth power.

$$\left(\sqrt[4]{x-1}\right)^4 = 3^4$$
$$x - 1 = 81$$
$$x = 82$$

This also works in reverse. In other words, you can cancel out an exponent n by taking the nth root. For example, to solve the equation $5x^3 = 80$, isolate the quantity raised to the exponent (just like you must isolate the radical to solve a radical equation).

$$\frac{5x^3}{5} = \frac{80}{5}$$
$$x^3 = 16$$

To cancel out the exponent, take the third root of both sides of the equation.

$$\sqrt[3]{x^3} = \sqrt[3]{16}$$
$$x = \sqrt[3]{8 \cdot 2}$$
$$x = \sqrt[3]{2^3 \cdot 2}$$
$$x = 2\sqrt[3]{2}$$

One thing to keep in mind when canceling out that exponent: if it's even, stick in a "±" sign on the left side of the equation when the radical signs appear. For example, to cancel out the even power of the equation $x^2 = 16$, insert a "±" sign when you take the square root of both sides of the equation.

$$\sqrt{x^2} = \pm\sqrt{16}$$
$$x = \pm 4$$

Bugging Out with Squares

When factoring fails, you need another option, an alternative way to solve the quadratic equation. One strong option is completing the square. A brief warning before you start: this problem contains a scatological reference, and is therefore rated PG by the National Advisory Council on Lowbrow Humor in Mathematics.

Example 2: Solve the equation $2x^2 - 16x + 10 = 0$ by completing the square.

Solution: You can't factor the trinomial into two linear binomials like you could in Examples 1(a) and 1(b) a few pages ago, so you should complete the square.

Kelley's Cautions

If you don't force the coefficient of x^2 to be 1, you'll get stuck in Step 4.

Step 1: Make sure the coefficient of x^2 is 1. If it's not, divide everything in the equation by that coefficient. In this problem, the coefficient of x^2 is 2, so divide everything by 2.

$$\frac{2x^2}{2} - \frac{16x}{2} + \frac{10}{2} = \frac{0}{2}$$
$$x^2 - 8x + 5 = 0$$

Step 2: Move everything to the left side of the equation except the constant. You want variable terms on the left and a constant sitting alone on the right side of the equation. In this example, that means subtracting 5 from both sides.

$$x^2 - 8x = -5$$

Step 3: Inspect bug droppings. This is the key step. You're going to add a mystery number to both sides of the equation. To figure out what number you should add, take half of the x-coefficient $\left(\frac{1}{2}(-8)=-4\right)$ and square it $((-4)^2 = 16)$. In this case, you should add 16 to both sides of the equation.

I don't know when, why, or how I came up with it, but for years I have used a little, three-segmented bug (pictured in Figure 13.1) to help me with step three. To this day, I still draw his body (minus the legs, eyes, and other time-consuming details) in the margins when I complete the square.

Figure 13.1

The bug ingests the coefficient of the x-term (–8), eats half of it (–8 ÷ 2 = –4), and leaves a piece of "square waste" (4² = 16).

Here's how my little invertebrate mnemonic friend works: the bug eats the x-coefficient (place the x-coefficient, including its sign, in the top segment). However, he's a small bug with a small appetite, only able to eat half of that number (put half of the number from the top segment, including its sign, in the second segment). Finally, once that number's eaten, he must, ahem, excrete some waste to complete the digestive process, and this bug leaves only *square* droppings (the square of the middle segment is written in the bug's rear end). Add that number to both sides of the equation.

$$x^2 - 8x = -5$$
$$x^2 - 8x + 16 = -5 + 16$$
$$x^2 - 8x + 16 = 11$$

Step 4: Rewrite the left side of the equation as a perfect square. When factored, the quadratic on the left side will be $(x + a)^2$. Here's the cool part: the a in $(x + a)^2$ comes right from the stomach of the bug, so in this case $a = -4$. Of course, you can factor the trinomial without using the bug's midsection (it's just regular, run-of-the-mill factoring) but this is a great shortcut when fractions and other uglier values are involved.

$$(x - 4)^2 = 11$$

Step 5: Eliminate the exponent. Take the square root of both sides of the equation; don't forget to stick a "±" sign on the right side.

$$\sqrt{(x-4)^2} = \pm\sqrt{11}$$
$$x - 4 = \pm\sqrt{11}$$

Step 6: Solve for x. In this problem, that means adding 4 to both sides of the equation.

$$x = 4 \pm \sqrt{11}$$

Therefore, the quadratic equation $2x^2 - 16x + 10 = 0$ has two solutions: $x = 4 + \sqrt{11}$ and $x = 4 - \sqrt{11}$.

You've Got Problems

Problem 2: Solve the equation $x^2 + 6x - 3 = 0$ by completing the square.

The Quadratic Formula

The final method you can use to solve quadratic equations is called (appropriately enough) the quadratic formula. Like completing the square, it works for all quadratic equations, and it's very easy to use. If you set a quadratic equation equal to 0, it will look like this: $ax^2 + bx + c = 0$ (where a, b, and c are real numbers). To solve the equation, plug a, b, and c into the quadratic formula:

$$x = \frac{-b \pm \sqrt{b^2 - 4ac}}{2a}$$

That formula may look ugly, but you need to memorize it. If you need help doing that, check out my web page www.calculus-help.com and click on the "Fun Stuff" section. I wrote a little song (rated PG for comedic violence) to help you remember it.

Example 3: Solve the equation $2x^2 - 5x = -1$ using the quadratic formula.

Solution: Set the equation equal to 0 by adding 1 to both sides.

$$2x^2 - 5x + 1 = 0$$

The equation is now in the form $ax^2 + bx + c = 0$ when $a = 2$, $b = -5$, and $c = 1$. Plug those values into the quadratic formula.

$$x = \frac{-b \pm \sqrt{b^2 - 4ac}}{2a}$$

$$x = \frac{-(-5) \pm \sqrt{(-5)^2 - 4(2)(1)}}{2(2)}$$

$$x = \frac{5 \pm \sqrt{25 - 8}}{4}$$

$$x = \frac{5 \pm \sqrt{17}}{4}$$

How'd They Do That?
Wondering where the quadratic formula comes from? Is it, too, the product of insect waste? Actually, in a manner of speaking, it is. The quadratic formula is the solution to the equation $ax^2 + bx + c = 0$ when solved by completing the square.

The solution to the quadratic equation $2x^2 - 5x = -1$ is $x = \dfrac{5 + \sqrt{17}}{4}$ or $x = \dfrac{5 - \sqrt{17}}{4}$. You could rewrite the solution as $x = \dfrac{5}{4} + \dfrac{\sqrt{17}}{4}$ or $x = \dfrac{5}{4} - \dfrac{\sqrt{17}}{4}$, but none of those fractions can be simplified, so there's no need to.

You've Got Problems
Problem 3: Solve the equation $x^2 + 6x - 3 = 0$ from Problem 2 using the quadratic formula, and verify that you get the same answer.

All Signs Point to the Discriminant

Have you ever owned one of those Magic 8 Balls? They look like comically oversized pool balls, but have a flat window built into them so that you can see what's inside—a 20-sided die floating in opaque blue goo. Supposedly, the billiard ball has prognostic powers; ask it a question, give it a shake, and slowly, mystically, like a petroleum-covered seal emerging from an oil spill, the die will rise to the little window and reveal the answer to your question.

The quadratic equation contains a Magic 8 Ball of sorts. The expression $b^2 - 4ac$ from beneath the radical sign is called the *discriminant*, and it tells you how many solutions a quadratic equation has, without the need for blue prognostication goo.

Talk the Talk
The **discriminant** is the expression $b^2 - 4ac$, defined for the quadratic equation $ax^2 + bx + c = 0$. The sign of the discriminant tells you how many real number solutions the quadratic equation has.

A lot of work is required to solve quadratic equations that you can't factor—tons of arithmetic abounds in the quadratic formula, and a whole bunch of steps are required to complete the square. Therefore, it's useful to gaze into the mystic beyond to make sure a quadratic equation even *has* any real number solutions before you spend any time trying to find them.

Here's how the discriminant works. Given a quadratic equation $ax^2 + bx + c = 0$, plug the coefficients into the expression $b^2 - 4ac$. Pay close attention to the sign of the number that results:

- If you get a positive number, the quadratic has two unique solutions.

- If you get 0, the quadratic has exactly one solution, a double root.

- If you get a negative number, the quadratic has no real solutions, just two complex ones (because they contain the value $i = \sqrt{-1}$ discussed in Chapter 12.)

The discriminant isn't magic. It just shows how important the role of $\sqrt{b^2 - 4ac}$ in the quadratic formula really is. Look at what happens when the radicand is 0:

$$x = \frac{-b \pm \sqrt{b^2 - 4ac}}{2a} = \frac{-b \pm \sqrt{0}}{2a} = -\frac{b}{2a}$$

The $\pm$ vanishes, taking with it the possibility of a second solution and thereby guaranteeing a double root. If, however, $b^2 - 4ac$ is negative, that means there's a negative inside a square root, indicating only imaginary solutions.

Example 4: Without calculating them, determine how many real solutions the equation $3x^2 - 2x = -1$ has.

Solution: Set the equation equal to 0 by adding 1 to both sides.

$$3x^2 - 2x + 1 = 0$$

Set $a = 3$, $b = -2$, and $c = 1$, and evaluate the discriminant.

$$b^2 - 4ac = (-2)^2 - 4(3)(1)$$
$$= 4 - 12$$
$$= -8$$

Because the discriminant is negative, the quadratic equation has no real number solutions, only two complex ones.

You've Got Problems

Problem 4: Without calculating them, determine how many real solutions the equation $25x^2 - 40x + 16 = 0$ has.

Solving One-Variable Quadratic Inequalities

According to Chapter 7, the solution to a simple inequality is expressed as a graph on a number line. This is still true of quadratic inequalities like $2x^2 + x - 2 < 0$.

Why do you use a number line instead of a coordinate plane? The number of axes in the graphing system needs to match the number of unique variables in the inequality. Because $2x^2 + x - 2 < 0$ only contains x (not x and y), the one-axis number line is used instead of the two-axis coordinate plane.

Enough talk for now; here's how to solve and graph a quadratic inequality:

Talk the Talk

Critical numbers are values of x for which an inequality either equals 0 or is undefined. They break the number line into segments called **intervals**.

1. **Pretend the inequality symbol is an equal sign and solve the corresponding quadratic equation.** Those solutions are the *critical numbers* of the inequality. You're not changing that inequality sign to an equal sign permanently—once you find the critical numbers, change it back.

2. **Graph the solutions on a number line.** Mark the critical numbers on the number line with either open or closed dots, depending on whether or not the symbol allows for the possibility of equality. (Use open dots for < or >, and use closed dots for ≤ or ≥.)

3. **Use test values to find the solution interval(s).** The critical numbers will split the number line into segments called *intervals*. Choose a test value from each interval and plug it into the original inequality for x. If the test value makes the inequality true, then so will *all* of the other values in that interval, so it represents at least part of the solution. (More than one interval may make the inequality true.)

4. **Graph the inequality.** Darken the segments of the number line that correspond to each solution interval.

Even though the graph you create during this process tells you what the solution is, most instructors want more than just a graph. They want an answer expressed as an inequality statement.

Example 5: Solve the inequality $2x^2 + x - 2 < 0$ and graph the solution.

Solution: Pretend, for a moment, that this is actually the equation $2x^2 + x - 2 = 0$. Unfortunately, this quadratic cannot be factored, so you'll have to use either the quadratic formula or complete the square to get the solutions, which will be:

$$x = -\frac{1}{4} - \frac{\sqrt{17}}{4} \quad \text{or} \quad x = -\frac{1}{4} + \frac{\sqrt{17}}{4}$$

Those critical numbers sure are ugly, so it's a good idea to type them into a calculator to figure out what decimal they're approximately equal to. That way, you can plot them on the number line like in Figure 13.2. The inequality symbol $<$ does not allow equality, so use open dots.

$$x = -\frac{1}{4} - \frac{\sqrt{17}}{4} \approx -1.28$$

$$x = -\frac{1}{4} + \frac{\sqrt{17}}{4} \approx 0.78$$

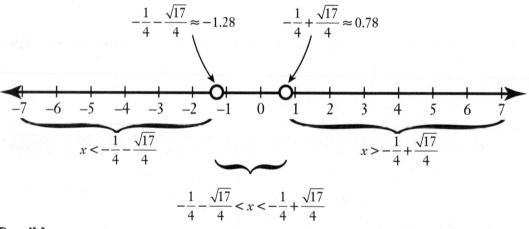

Figure 13.2

The critical numbers (approximately −1.28 and 0.78) break the number line into three intervals.

Choose one value to represent each interval (like $x = -2$, $x = 0$, and $x = 1$) and plug them into the original inequality to see which values produce true statements.

Test $x=-2$	Test $x=0$	Test $x=1$
$2(-2)^2+(-2)-2<0$	$2(0)^2+(0)-2<0$	$2(1)^2+(1)-2<0$
$2(4)-2-2<0$	$2(0)-2<0$	$2(1)+1-2<0$
$8-4<0$	$0-2<0$	$2-1<0$
$4<0$ False	$-2<0$ True	$1<0$ False

Only the interval containing $x = 0$ (the middle interval in Figure 13.2) makes the inequality true, so it represents the solution:

$$-\frac{1}{4}-\frac{\sqrt{17}}{4}<x<-\frac{1}{4}+\frac{\sqrt{17}}{4}$$

To graph the solution, darken the solution interval on the number line in Figure 13.2; you'll end up with Figure 13.3.

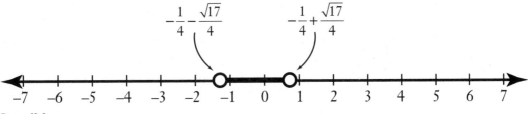

Figure 13.3

The solution graph for the inequality $2x^2 + x - 2 < 0$.

You've Got Problems

Problem 5: Solve the inequality $2x^2 + 5x - 3 \geq 0$ and graph the solution.

The Least You Need to Know

- When you complete the square, the leading coefficient must be 1.

- Completing the square requires you to take half of the x-coefficient and square it; this number should be added to both sides of the equation.

- The quadratic formula is $x = \dfrac{-b \pm \sqrt{b^2 - 4ac}}{2a}$.

- The discriminant $b^2 - 4ac$ is used to determine how many solutions a quadratic equation will have.

14

Solving High-Powered Equations

In This Chapter

- ◆ Solving cubic and higher-degree equations
- ◆ Factoring large polynomials
- ◆ Calculating rational and irrational roots

As you found out in Chapter 13, the methods used to solve linear equations and quadratic equations are dramatically different. Are you asking yourself, "Will I have to learn how to solve equations all over again every time the degree increases by 1?" or "Will I have to learn a new language (something as weird and antiquated as Latin or perhaps as complicated as dolphin) to figure these things out?"

I have good news for you; the answer to both of those questions is "no." The techniques you use to solve third-degree equations are the same ones you use to solve fourth-, fifth-, sixth-, and fiftieth-degree equations, not that you'll see many of those. In fact, you can even use these techniques to go back and solve linear and quadratic equations (if you're feeling retro).

There's No Escaping Your Roots!

Before you tackle equations with large exponents, think about the way you solve quadratic equations by factoring. Let's say you have some quadratic equation set equal to 0, and you were able to factor it like this: $(x - a)(x - b) = 0$, where a and b are real numbers. Remember what to do next? Set each factor equal to 0 and solve those equations to find the solutions (which you can also call the *roots* of the equation).

$$x - a = 0 \qquad x - b = 0$$
$$x = a \quad \text{or} \quad x = b$$

This example illustrates an important relationship between the factors of a polynomial and its roots: if $x - a$ is a factor of a polynomial, then $x = a$ is a solution, or root, when that polynomial is set equal to 0.

Minus all the techno mumbo jumbo, what does that mean for you? To find the solutions of a polynomial equation set equal to 0, set its factors equal to 0 and solve. Occasionally, when factoring is impossible, you'll have to resort to the quadratic formula.

Talk the Talk

The solutions of an equation are also called the roots.

That's the game plan. However, there's one problem. You probably know how to factor something like $x^2 - 4x + 3$, but factoring $x^3 - 6x^2 + 5x + 12$ is a different matter entirely! If higher powers are involved (here I mean exponents, not divine intervention), how are you supposed to factor? Good question. Keep reading and all will be revealed.

Finding Factors

Remember, a factor is something that divides evenly into something else. In other words, if x is a factor of y, then the quotient $y \div x$ has a remainder of 0. Factors work the same way with polynomials. If $x - m$ is a factor of the polynomial $ax^4 + bx^3 + cx^2 + dx + e$, then the quotient $(ax^4 + bx^3 + cx^2 + dx + e) \div (x - m)$ has a remainder of 0.

Critical Point

If your synthetic division skills are a little rusty, flip back to Chapter 10 to review. Remember, the rightmost number beneath the horizontal line is the remainder.

Example 1: Demonstrate that $x - 3$ is a factor of the polynomial $x^3 - 6x^2 + 5x + 12$ and use that information to completely factor the cubic.

Solution: If $x - 3$ is a factor of $x^3 - 6x^2 + 5x + 12$, then dividing that cubic expression by $x - 3$ should produce a remainder of 0. Because $x - 3$ is linear, you can use synthetic division instead of the more tedious process of long division. Don't forget that you should write 3, not -3, in the synthetic division box.

Critical Point _____

By determining whether or not $x - 3$ is a factor of $x^3 - 6x^2 + 5x + 12$, you are simultaneously determining whether or not 3 is a root of the equation $x^3 - 6x^2 + 5x + 12 = 0$.

$$
\begin{array}{r|rrrr}
3 & 1 & -6 & 5 & 12 \\
  &   & 3 & -9 & -12 \\
\hline
  & 1 & -3 & -4 & 0
\end{array}
$$

The remainder is 0, so $x - 3$ is a factor, and you can rewrite the polynomial in factored form: $(x - 3)(x^2 - 3x - 4)$. Notice that the quadratic that's left can also be factored: $x^2 - 3x - 4 = (x - 4)(x + 1)$. Therefore, the fully factored version of $x^3 - 6x^2 + 5x + 12$ is $(x - 3)(x - 4)(x + 1)$. The order of the factors doesn't matter, thanks to the commutative property of multiplication.

You've Got Problems

Problem 1: Demonstrate that $x + 2$ is a factor of the polynomial $2x^3 + 5x^2 - 8x - 20$ and use that information to completely factor the trinomial.

Baby Steps to Solving Cubic Equations

Have you ever seen the movie *What About Bob?* starring Bill Murray and Richard Dreyfuss? Murray is a patient under the psychiatric care of mental health guru Dreyfuss, and greatly benefits from the doctor's philosophy of "baby steps." Paraphrased, this philosophy entails looking at a complex problem as a series of small, easily accomplished (baby) steps, rather than one giant leap from start to finish.

In keeping with this fictional, but useful, therapeutic technique, I have inserted a baby step into the process of solving high-powered equations. Example 1 asked you to factor a polynomial, but in Example 2, you'll factor the polynomial and then take a baby step, solving an equation containing that polynomial. There's not a lot of difference between the two goals, but if not for this baby step, you might trip over the next topic, so trust me on this one.

Critical Point _____

A polynomial equation has up to n rational roots, where n is the degree of the polynomial. It may have fewer, but it cannot have more.

As you solve these problems, cement in your mind the relationship between factors and roots within a high-powered equation. Remember, if a is a root of a polynomial equation set equal to 0, then $x - a$ is one of the polynomial's factors—the sign of a changes when you convert from a root to a factor, and vice versa.

Example 2: If 5 is a root of the equation $x^3 + x^2 - 22x - 40 = 0$, what are the other two roots?

Solution: If 5 is a root of the equation, then $x - 5$ must be a factor. Use synthetic division to calculate the quotient $(x^3 + x^2 - 22x - 40) \div (x - 5)$. Chapter 10 told you to put a in the synthetic division box when you're factoring out (or dividing by) $x - a$, and now you know why: you're actually writing the corresponding root in there.

$$\begin{array}{r|rrrr} 5 & 1 & 1 & -22 & -40 \\ & & 5 & 30 & 40 \\ \hline & 1 & 6 & 8 & 0 \end{array}$$

Therefore, $x^3 + x^2 - 22x - 40 = (x - 5)(x^2 + 6x + 8)$. Factor the quadratic that's left and write the original cubic equation in its fully factored form.

$$(x - 5)(x + 4)(x + 2) = 0$$

Set each factor equal to 0 and solve those equations.

$$\begin{array}{ccccc} x - 5 = 0 & & x + 4 = 0 & & x + 2 = 0 \\ & \text{or} & & \text{or} & \\ x = 5 & & x = -4 & & x = -2 \end{array}$$

You've Got Problems

Problem 2: If -2 is a root of the equation $x^3 + 10x^2 + 7x - 18 = 0$, what are the other two roots?

Calculating Rational Roots

Does it feel like you've been cheating a little bit? The goal of the chapter is to solve, or find the roots of, high-powered equations, and in every example so far, I have given you one of the roots or factors and asked you to find the others based on it. What happens when no one hands you one of the answers, providing the kick start you need, that seed number for synthetic division that breaks the bigger, uglier polynomial into something like a quadratic that you can factor on your own?

Enter the heroic and handy *rational root test*, which generates a list of all the numbers that could *possibly* be the rational roots of the equation. That's important because the roots you'll use to factor need to be rational in order to apply synthetic division.

Talk the Talk _____

The **rational root test** uses the leading coefficient and constant of a polynomial to generate a list of possible rational root for a polynomial equation. Unfortunately, the list is usually much longer than the actual list of rational roots, so some trial and error is necessary to identify the actual roots.

Basically, the rational root test is like a criminal lineup at the local police department. If you're a witness to a crime, the police may round up a bunch of thugs that loosely match your description of the guilty party and ask you to identify the actual criminal from among the people in that lineup. Not all of the people there are guilty of the crime (at least not the crime you witnessed, anyway), so it's your job to identify the responsible party or parties in the much larger group of possible suspects.

Back to the (slightly more boring) world of math. To round up all the possible rational root suspects of an equation, the rational root test looks at the leading coefficient and the constant of a polynomial. More specifically, it looks at the factors of those numbers. It then lists all possible fractions of the form $\pm\dfrac{C}{L}$, where C is a factor of the constant and L is a factor of the leading coefficient. Sound strange? Wait until you see how it works in Example 3—it's not so bad.

After all of the possible suspects are identified, it's time to separate the guilty (roots) from the innocent (nonroots). Use synthetic division, giving each potential root a turn inside the synthetic division box to see which results in a remainder of 0. Once you spot that telltale remainder, tell the local police chief to slap on the cuffs, and then go back to the lineup, until all of the guilty suspects are apprehended. The degree of the polynomial indicates the maximum number of suspects, so if a polynomial has degree three, it has as many as three rational roots.

Example 3: Identify all the rational roots for each equation.

(a) $2x^3 - 9x^2 - 8x + 15 = 0$

Solution: List all the factors of the leading coefficient (2) and the constant (15).

Factors of 2:   1, 2

Factors of 15:   1, 3, 5, 15

The rational root test tells you that all the positive and negative fractions you can create with numerator 1, 3, 5, or 15 and denominator 1 or 2 are possible rational roots of the equation. The easiest way to make the list is to take each possible numerator and write it over each possible denominator.

$$\pm\frac{1}{1}, \pm\frac{1}{2}, \pm\frac{3}{1}, \pm\frac{3}{2}, \pm\frac{5}{1}, \pm\frac{5}{2}, \pm\frac{15}{1}, \pm\frac{15}{2}$$

Critical Point _____

You won't always hit an actual root on your first or second stab at synthetic division, but stick with it until you find that elusive zero remainder. If your instructor allows you to use graphing calculators, here's a trick: test roots that appear to be x-intercepts of the polynomial's graph.

Yikes! There are 16 possible suspects (8 positive and 8 negative), and only a few of them are actually guilty of the crime of Criminal Activity in the Third Degree (Equation). Try them in a court of synthetic division, one at a time, until you get a remainder of 0. You might as well start with –1, for no earth-shattering reason other than it's at the beginning of the suspect list.

$$
\begin{array}{r|rrrr}
-1 & 2 & -9 & -8 & 15 \\
   &   & -2 & 11 & -3 \\
\hline
   & 2 & -11 & 3 & 12 \\
\end{array}
$$

Nuts, that didn't work because the remainder was 12 (instead of 0). Oh well, one suspect cleared and 15 to go. This time, test to see if its opposite, 1, is a dirty, rotten root.

$$
\begin{array}{r|rrrr}
1 & 2 & -9 & -8 & 15 \\
  &   & 2 & -7 & -15 \\
\hline
  & 2 & -7 & -15 & 0 \\
\end{array}
$$

Success! If 1 is a root, that means x 1 is a factor.

$$2x^3 - 9x^2 - 8x + 15 = 0$$
$$(x-1)(2x^2 - 7x - 15) = 0$$

Factor the leftover quadratic expression.

$$(x - 1)(2x + 3)(x - 5) = 0$$

Set each factor equal to 0 and solve.

$$x - 1 = 0 \quad \text{or} \quad 2x + 3 = 0 \quad \text{or} \quad x - 5 = 0$$
$$x = 1 \qquad x = -\frac{3}{2} \qquad x = 5$$

The roots of the equation are $-\dfrac{3}{2}$, 1, and 5.

(b) $x^4 - 3x^3 - 19x^2 + 27x + 90 = 0$

Solution: The leading coefficient only has a factor of 1, but the constant's factors are 1, 2, 3, 5, 6, 9, 10, 15, 18, 30, 45, and 90. So, the possible rational roots are:

$$\pm 1, \pm 2, \pm 3, \pm 5, \pm 6, \pm 9, \pm 10, \pm 15, \pm 18, \pm 30, \pm 45, \text{ and } \pm 90$$

(Each possible root has a denominator of 1.) Once again, start at the beginning of the list. If you synthetically divide –1, 1, –2, and 2, you get nonzero remainders, but once you get to –3, your luck takes a turn for the better.

$$
\begin{array}{r|rrrrr}
-3 & 1 & -3 & -19 & 27 & 90 \\
   &   & -3 & 18 & 3 & -90 \\
\hline
   & 1 & -6 & -1 & 30 & 0
\end{array}
$$

Therefore, the polynomial can be factored as $(x + 3)(x^3 - 6x^2 - x + 30)$, but an ugly trinomial remains, not an easily factored binomial like in previous examples. Oh well, back to the list of suspects. You don't have to try any of the roots that already failed (they still won't work), but when you test other possible roots, use $x^3 - 6x^2 - x + 30$ instead of $x^4 - 3x^3 - 19x^2 + 27x + 90$. It turns out that 3 is also a root.

> **CAUTION**
>
> **Kelley's Cautions**
>
> If you don't use the one-degree-smaller result of synthetic division each time you find a root (rather than the original polynomial) like in Example 3(b), you'll never whittle the polynomial down into a quadratic you can factor.

$$
\begin{array}{r|rrrr}
3 & 1 & -6 & -1 & 30 \\
  &   & 3 & -9 & -30 \\
\hline
  & 1 & -3 & -10 & 0
\end{array}
$$

Once you factor $x + 3$ and $x - 3$ out of the quartic, you're left with $x^2 - 3x - 10$.

$$x^4 - 3x^3 - 19x^2 + 27x + 90 = 0$$
$$(x + 3)(x - 3)(x^2 - 3x - 10) = 0$$

Factor the quadratic that's left and solve.

$$(x+3)(x-3)(x-5)(x+2)=0$$

$$x+3=0 \qquad x-3=0 \qquad x-5=0 \qquad x+2=0$$
$$x=-3 \qquad x=3 \qquad x=5 \qquad x=-2$$

The roots of the equation are –3, –2, 3, and 5.

You've Got Problems
Problem 3: Identify all four rational roots of the polynomial $4x^4 + 4x^3 - 9x^2 - x + 2 = 0$.

What About Imaginary and Irrational Roots?

So far, these high-powered equations have worked out nicely. In the end, you're always getting a total number of rational roots that's equal to the degree of the polynomial, so you might be tempted to assume that will always be true. Unfortunately, it's not. Occasionally, you'll end up with some irrational roots (that contain radicals which simply refuse to be simplified) or imaginary roots.

How'd They Do That?
According to the Fundamental Theorem of Algebra, a polynomial of degree n, when set equal to 0, has exactly n roots. They may be real, imaginary, rational, or irrational, but their number will always match the degree of the equation. Note that imaginary roots will always come in pairs (that are conjugates of one another) and a double root counts as two roots.

You'll just be tooling along, trying to locate rational roots, breaking polynomials up into smaller factors using synthetic division, and all of a sudden you'll end up with a quadratic that just doesn't factor. No problem—apply the quadratic formula and simplify.

Example 4: Identify all roots of each equation.

(a) $x^3 - 8x^2 + 4x - 32 = 0$

Solution: Use the rational root test to generate a list of possible rational root suspects:

$$\pm 1, \pm 2, \pm 4, \pm 8, \pm 16, \pm 32$$

Believe it or not, the only rational root is 8.

$$
\begin{array}{r|rrrr}
8) & 1 & -8 & 4 & -32 \\
   &   & 8 & 0 & 32 \\
\hline
   & 1 & 0 & 4 & 0
\end{array}
$$

Factor the equation to get $(x - 8)(x^2 + 4) = 0$. Notice that $x^2 + 4$ is the *sum* of perfect squares (not the *difference* of perfect squares), and therefore, it cannot be factored. However, you can set that factor equal to 0 and solve the equation.

$$x^2 + 4 = 0$$
$$x^2 = -4$$
$$\sqrt{x^2} = \pm\sqrt{-4}$$
$$x = +2i$$

The roots of the equation are 8, $2i$, and $-2i$. Notice that there are three roots (which matches the degree of the polynomial), and the imaginary roots are conjugates.

(b) $x^4 + 2x^3 - 10x^2 - 5x + 12 = 0$

Solution: Of the possible rational roots ($\pm 1, \pm 2, \pm 3, \pm 4, \pm 6$, and ± 12) only -4 and 1 are valid. After you apply synthetic division for both roots, you end up with $(x - 1)(x + 4)(x^2 - x - 3) = 0$. The quadratic expression is prime, so when you set it equal to 0, apply the quadratic formula.

$$x = \frac{-(-1) \pm \sqrt{(-1)^2 - 4(1)(-3)}}{2(1)} = \frac{1 \pm \sqrt{1+12}}{2} = \frac{1 \pm \sqrt{13}}{2}$$

The roots of the equation are -4, 1, $\frac{1}{2} - \frac{\sqrt{13}}{2}$, and $\frac{1}{2} + \frac{\sqrt{13}}{2}$. Once again, the total number of roots (two rational and two irrational for a total of four) matches the degree of the polynomial.

You've Got Problems

Problem 4: Identify all roots of the equation $2x^3 - 5x^2 + 4x - 21 = 0$.

The Least You Need to Know

◆ The solutions of an equation are also called roots.

◆ If $x - b$ is a factor of a polynomial, then b is a root of the equation created by setting the polynomial equal to 0.

◆ The rational root test lists all possible rational roots of a polynomial equation based on its constant and leading coefficient.

◆ A polynomial equation of degree n always has exactly n roots, although some may be irrational or imaginary.

Part 5

The Function Junction

In his book *Prey*, Michael Crichton describes the horror of what can happen when nanomachines (tiny little robots the size of molecules) go out of control and wreak havoc in a murderous rampage. In this part, you learn about the nanomachines called functions that power mathematics. Luckily, they're only theoretical and pose no physical danger to you whatsoever, unless of course you drop your algebra book on your foot or something. Then they hurt like crazy.

15

Introducing the Function

In This Chapter

- ◆ Distinguishing between functions and relations
- ◆ Combining functions
- ◆ Creating inverse functions
- ◆ Evaluating piecewise-defined functions

Brace yourself, because it's time to change algebraic gears. Everything leading up to this moment has dealt with expressions and equations, and although you will continue to see both, you're going to see a lot more *functions*. They're a lot like the equations and expressions you're used to, but the characteristics that make them unique, while minor, have major consequences.

Getting to Know Your Relations

Before I explain to you what a function is, I need to step back for a moment and discuss vending machines. As a one-time college student, I fully appreciate the importance of these handy, rectangular, poor man's concierges.

The rickety but trustworthy dorm vending machine (helpfully labeled "Tasty Snacks" and centrally located in the TV lounge) didn't mind if it was 2 A.M. when you came looking for peanut butter cheese crackers. For a mere 50¢ and a press of the right button, you were seconds away from cellophane-wrapped peanut-buttery goodness (and inexplicably orange fingers).

In the mathematical world, there are an infinite number of little vending machines called *relations*, rules that swallow a number input (like a vending machine swallows a coin) and output another number (like a snack machine dispenses treats for your taste buds and cholesterol directly to the lining of your artery walls).

As you might guess, the job of a relation is to define a *relationship* between the inputs and outputs of the vending machine. In other words, a relation provides the answer to "What will you give me if I give you this?" The simplest way, therefore, to define a relation is to list the inputs and the corresponding outputs as ordered pairs, like so:

$$s: \{(1,3), (2,7), (3,-1), (4,9), (5,-1), (5,0)\}$$

This relation, called s, tells you that an input of 1 results in an output of 3, thanks to the coordinate pair (1,3). If you walk up to the s vending machine and pop a 1 in the coin slot, a 3 pops out (possibly coated in peanut butter and/or containing nougat). Mathematically you write $s(1) = 3$, which is read "s of 1 equals 3." Similarly, you could write $s(2) = 7$, $s(3) = -1$, and so forth.

Something buggy happens to the s vending machine when you input 5—did you notice? According to the definition of s, $s(5) = -1$ and $s(5) = 0$. In other words, if you plug 5 into the relation, you might get one answer, and you might get the other answer. You might even get both (if you shook the vending machine hard enough, I suppose).

Talk the Talk

A **relation** is a rule that accepts input values and produces corresponding output values. If every input has only one corresponding output, then the relation is classified as a **function**.

Some relations receive a special commendation for good and reliable service; those relations are called *functions*. However, s will never receive this commendation, because it does not meet the one and only requirement that makes a relation a function: every input must be paired with a single output. Because s fails that litmus test (the input 5 is paired with *two* possible outputs, −1 and 0), it is not a function.

Functions and relations are rarely written as lists of ordered pairs, because most have infinitely many inputs, and writing out an infinite list of numbers is not the best use of anyone's time. Instead, they're usually written in terms of the input, usually x, like this:

$$h(x) = 3x + 4$$

The relation h is also a function, because no matter what you substitute in for x, you'll always get a single, predictable, corresponding answer. For instance, the input $x = -3$ always produces an output of -5.

$$\begin{aligned} h(-3) &= 3(-3) + 4 \\ &= -9 + 4 \\ &= -5 \end{aligned}$$

No matter how many times you plug in $x = -3$, you'll get -5 every time, and that's true for any input.

How'd They Do That?

Consider the relation r defined below.

$$r: \{(-2,4), (-1,9), (0,4)\}$$

There is something troubling about r. Notice that both $r(-2)$ and $r(0)$ are equal to 4. In other words, if you input either -2 or 0, you'll get the same thing out: 4. So, is r a function? It is, because each input corresponds to only one output.

There's no rule that says you can't get the same output for multiple inputs. However, there is a special name for functions whose inputs each have a unique output—they are called *one-to-one functions*.

Don't worry about how to test relations to determine whether they're functions right now. In Chapter 16, you'll be introduced to the vertical line test, a quick way to figure out whether a relation is a function.

Operating on Functions

Every time a new algebraic concept pops up, the first things you usually learn are how to add, subtract, multiply, and divide those new things. (Think back to fractions, matrices, polynomials, and radicals, just to name a few.) However, because functions are made up of things like fractions, polynomials, and radicals, combining them is easy. The only truly new thing you'll learn is the notation to use as you combine the functions.

Example 1: Given the functions $f(x) = x^2 - 4x + 3$ and $g(x) = x - 1$, identify the following functions and evaluate each when $x = 2$.

(a) $(f + g)(x)$

Solution: Add the quadratic function $f(x)$ to the linear function $g(x)$ by combining like terms.

$$(f + g)(x) = (x^2 - 4x + 3) + (x - 1)$$
$$= x^2 + (-4x + x) + (3 - 1)$$
$$= x^2 - 3x + 2$$

Evaluate $(f + g)(2)$, as directed by the problem, by substituting $x = 2$ into $(f + g)(x)$.

$$(f + g)(2) = (2)^2 - 3(2) + 2$$
$$= 4 - 6 + 2$$
$$= 0$$

(b) $(f - g)(x)$

Solution: Distribute -1 through the quantity $x - 1$ before combining like terms.

$$(f - g)(x) = (x^2 - 4x + 3) - (x - 1)$$
$$= x^2 - 4x + 3 - x + 1$$
$$= x^2 - 5x + 4$$

Evaluate $(f - g)(2)$.

$$(f - g)(2) = 2^2 - 5(2) + 4$$
$$= 4 - 10 + 4$$
$$= -2$$

(c) $(fg)(x)$

Solution: Multiply the polynomials by multiplying each term of $f(x)$ by each term of $g(x)$, one at a time.

$$(fg)(x) = (x^2 - 4x + 3)(x - 1)$$
$$= x^2(x) + x^2(-1) + (-4x)(x) + (-4x)(-1) + 3(x) + 3(-1)$$
$$= x^3 - x^2 - 4x^2 + 4x + 3x - 3$$
$$= x^3 - 5x^2 + 7x - 3$$

Evaluate $(fg)(2)$.

$$(fg)(2) = 2^3 - 5(2)^2 + 7(2) - 3$$
$$= 8 - 20 + 14 - 3$$
$$= -1$$

(d) $\left(\dfrac{f}{g}\right)(x)$

Solution: Because $x - 1$ is a factor of $x^2 - 4x + 3$, you can simplify the quotient $\left(\dfrac{f}{g}\right)(x)$.

$$\left(\frac{f}{g}\right)(x) = \frac{x^2 - 4x + 3}{x - 1}$$

$$= \frac{(x-3)\,(x-1)}{x-1}$$

$$= x - 3$$

Evaluate $\left(\dfrac{f}{g}\right)(2)$.

$$\left(\frac{f}{g}\right)(2) = 2 - 3 = -1$$

Critical Point

The way I evaluate $f(2)$ at the end of each part of Example 1 is not the only way to do it. Take part (a) as an example. Instead of plugging 2 into the new function $(f + g)(x)$, you can plug 2 into each of the original functions, $f(x)$ and $g(x)$, and then just add the results together.

$$f(2) = 2^2 - 4(2) + 3$$
$$= 4 - 8 + 3$$
$$= -1$$

$$g(2) = 2 - 1$$
$$= 1$$

Therefore, $(f + g)(2) = f(2) + g(2) = -1 + 1 = 0$, which matches the value of $(f + g)(2)$ from Example 1(a).

You've Got Problems

Problem 1: Given the functions $h(x) = x^2 + 7x - 5$ and $k(x) = 4x - 3$, evaluate the following:

(a) $(h + k)(-1)$

(b) $(h - k)(-1)$

(c) $(hk)(-1)$

(d) $\left(\dfrac{h}{k}\right)(-1)$

Composition of Functions

During college, my friends, Matt and Chris, and I decided to hold a contest, the sort of intellectual competition that's only dreamed up by those residing in the hallowed halls of academia—a clash of wit and cunning designed while our minds were at peak functioning power. Basically, we wanted to see who could stuff the most Cheerios in his mouth before he gagged or asphyxiated by drawing a legion of tan breakfast Os into his lungs.

Chris won. I don't remember the absurd number of Cheerios he was able to jam into his maw (and I don't recommend trying this at home), but I admired his dogged determination not to lose the contest although nothing was at stake (except bragging rights, of course). Even today, I believe some Cheerios reside in his sinus cavities as a de facto trophy of the event, accidentally sucked in when he started gagging, eyes wide in fright with the sudden fear of dying in a way that would be made famous on the Internet.

Kelley's Cautions

If a function $f(x)$ contains multiple x's, and you're trying to create the composition $f(g(x))$, make sure to plug $g(x)$ into every one of the x-variables in $f(x)$.

If anything, the Cheerio Jam taught me two major lessons. First, college students really know how to "make their own fun"; and second, it's always fun to jam more into something than you usually do.

So far this chapter, you've substituted real numbers into functions—normal, bite-sized mouthfuls that the functions chewed up easily and spit out as outputs. Now, however, the inputs are going to get bigger.

Rather than single numbers, you're going to feed functions using other functions, in a weirdly cannibalistic process called *composition of functions.*

The intent to plug a function $g(x)$ into another function $f(x)$ is indicated by the notation $(f \circ g)(x)$, read "f composed with g of x" or "f circle g of x." That little round thing in there is, indeed, a circle—a new operator that joins the ranks of +, −, x, ·, and ÷ that basically means "plug the right function in for the left function's variable." Sometimes, function composition is written explicitly, like this: $f(g(x))$ (read "f of g of x"), sort of like $f(2)$, except instead of plugging 2 into the function, you plug in $g(x)$ instead.

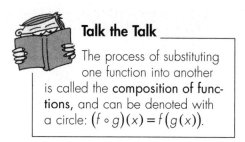

Talk the Talk

The process of substituting one function into another is called the **composition of functions**, and can be denoted with a circle: $(f \circ g)(x) = f(g(x))$.

Example 2: If $g(x) = 2x - 1$, and $h(x) = x^2 + 3x$, find $(h \circ g)(x)$.

Solution: The expression $(h \circ g)(x)$ means the same thing as $h(g(x))$, so plug $g(x)$, which equals $2x - 1$, into $h(x)$ wherever you see an x.

$$h(x) = x^2 + 3x$$

$$h(2x - 1) = (2x - 1)^2 + 3(2x - 1)$$

Notice that $h(x)$ contains two x-terms, so you've got to plug $2x - 1$ into both.

$$h(2x - 1) = 4x^2 - 4x + 1 + 6x - 3$$

$$= 4x^2 + 2x - 2$$

After you get over the initial unease of replacing a variable with something that has a variable in it, you're ready to start composing larger and larger functions, until you accidentally gag, drawing numbers and variables into your sinus cavity like my college roommate Chris.

You've Got Problems

Problem 2: If $f(x) = x^2 + 5$ and $g(x) = \sqrt{x - 2}$, identify the following functions:

(a) $(f \circ g)(x)$

(b) $(g \circ f)(x)$

Inverse Functions

According to Chapter 13, a root with index n cancels out the exponent n. For example, to solve the equation $(x - 2)^2 = 5$, you take the square root of both sides: $\sqrt{(x-2)^2} = \pm\sqrt{5}$. That is because squares and square roots are *inverse functions*.

Talk the Talk

Inverse functions cancel each other out if plugged into one another. In other words, $(f \circ g)(x) = (g \circ f)(x) = x$. The inverse of $f(x)$ is written $f^{-1}(x)$.

Inverse functions balance one another out in the mathematical world; they represent the eternal struggle between yin and yang, push and pull, right and left, really fun (like Disney's Magic Kingdom and MGM Studios in Florida) and extremely boring (like Disney's Epcot Center—"Hey, who wants to watch another 45-minute movie on the troposphere?").

Defining Inverse Functions

Mathematically speaking, if you compose inverse functions with one another, they cancel each other out. In other words, if $f(x)$ and $g(x)$ are inverses of one another, then:

$$f(g(x)) = g(f(x)) = x$$

See how they canceled each other out? If you plug $f(x)$ into $g(x)$ or $g(x)$ into $f(x)$, you get the same thing: x (the input of the innermost function). To denote that $g(x)$ is the inverse of $f(x)$, you write $g(x) = f^{-1}(x)$ (read "g of x is equal to f inverse of x," or "g of x is the inverse of f of x"). You can also write $f(x) = g^{-1}(x)$. Even though inverse function notation *looks* like a negative exponent, it's not, so don't start thinking about taking the reciprocal of $f(x)$ or anything.

Critical Point

Only one-to-one functions (whose inputs correspond to unique outputs) have inverses. For now, when I ask you to find the inverse of a function, you can assume that the function is one-to-one. Chapter 16 covers a technique that helps you determine whether or not a function is one-to-one.

Inverse functions also have another cool property: if you reverse the ordered pair in one function, you get an ordered pair for the inverse function. Therefore, if $f(a) = b$, then $f^{-1}(b) = a$. If that doesn't make sense to you, think of it this way. If you plug the number 2 into some function $f(x)$ and get out -7, then plugging -7 into $f^{-1}(x)$ gives you 2.

Example 3: Demonstrate that $f(x) = 4x + 3$ and $g(x) = \dfrac{1}{4}x - \dfrac{3}{4}$ are inverses functions.

Solution: If $f(x)$ and $g(x)$ are inverses, then $f(g(x))$ and $g(f(x))$ should both equal x.

$$f\left(g(x)\right) = f\left(\frac{1}{4}x - \frac{3}{4}\right)$$

$$g\left(f(x)\right) = g(4x+3)$$

$$= 4\left(\frac{1}{4}x - \frac{3}{4}\right) + 3$$

$$= \frac{1}{4}(4x+3) - \frac{3}{4}$$

$$= \frac{4}{4}x - \frac{12}{4} + 3$$

$$= \frac{4}{4}x + \frac{3}{4} - \frac{3}{4}$$

$$= x - 3 + 3$$

$$= x$$

$$= x$$

You've Got Problems

Problem 3: Demonstrate that $f(x) = 3x - 5$ and $g(x) = \dfrac{x+5}{3}$ are inverse functions.

Calculating Inverse Functions

Creating the inverse function $f^{-1}(x)$ for some function $f(x)$ is pretty simple—just follow these steps:

1. **Rewrite $f(x)$ as y.** This introduces a second variable into the equation.

2. **Switch x and y.** Wherever you see an x, change it to y and vice versa. This is how you force the function into an inverse. Remember, a function and its inverse contain reversed ordered pairs—if $f(x)$ contains the point (a,b), then $f^{-1}(x)$ contains (b,a).

3. **Solve for y.** Isolate y on one side of the equation.

4. **Rewrite y as $f^{-1}(x)$.** You're all finished. Go make a sandwich and relax.

That's all I'm going to say about inverse functions. Now that you can find them for yourself and verify that they are inverses, you've done enough for now. There's more to learn about inverse functions (including technical details such as domain restrictions) that are addressed by later courses, but I omit them here because discussing them now is unnecessarily complicated.

Example 4: If $f(x) = \sqrt{x+2}$, identify $f^{-1}(x)$.

Solution: Rewrite $f(x)$ as y.

$$y = \sqrt{x+2}$$

Switch the x's and y's.

$$x = \sqrt{y+2}$$

Solve for y. To eliminate the radical sign, square both sides of the equation. (You don't add a "±" sign—you only do that when you take the square *root* of both sides of an equation.)

$$x^2 = y + 2$$
$$x^2 - 2 = y$$

Rewrite y as $f^{-1}(x)$. You should also reverse the sides of the equation to get $f^{-1}(x)$ on the left side.

$$f^{-1}(x) = x^2 - 2$$

You've Got Problems

Problem 4: If $g(x) = 7x - 3$, identify $g^{-1}(x)$.

Bet You Can't Solve Just One: Piecewise-Defined Functions

I dated a few people before I met my wife. When I was very young, I figured that as soon as I found somebody that I liked (and who liked me), I'd get married. Little did I know that dating came first, and it was a lot more complicated than I'd figured—an exercise in determining what you can and can't put up with in a partner.

Consider your own dating history. One person's very nice, but too wishy-washy. Another person's very self-assured but too possessive. Yet another person has a repulsive personality, but doesn't mind the way your feet smell. Basically, the entire time you're dating, you're really designing a wish list for the perfect mate: he or she has to be self-assured, but not pushy; nice but not saccharinely sweet; and also able to put up with the smell of your toes when you kick off wet shoes. Luckily, someone eventually comes around who meets just about all of those qualities (face it, your feet aren't going to win you any prizes), and you take the plunge.

Things are a little different with functions; you can actually design the perfect function—one that meets any qualifications you set. In fact, if you like certain parts of different functions, you can take the parts you like, sew them together, and design a new function that's a patchwork quilt of your favorite function characteristics called a *piecewise-defined function*.

The small component functions that, together, comprise your dream function are bounded by one big brace on the left and look something like this:

$$f(x) = \begin{cases} g(x), x \le a \\ h(x), x > a \end{cases}$$

Talk the Talk

A piecewise-defined function is made up of two or more functions that are restricted according to input. Each individual function that makes up the piecewise-defined function is valid for only certain *x*-values.

The function $f(x)$ is actually made up of two other functions, $g(x)$ and $h(x)$. You'll use $g(x)$ whenever the input is less than or equal to a, but turn around and use the other function $h(x)$ whenever the input is greater than a.

Kelley's Cautions

You're not limited to two component functions inside that brace. As long as the input restrictions (like $x \le a$ and $x > a$) don't overlap, the result is a function. For instance, consider $h(x)$:

$$h(x) = \begin{cases} x^2 - 3, x < 4 \\ 2x + 1, x \ge 0 \end{cases}$$

Notice that an input of $x = 1$ can be plugged into both functions, because if $x = 1$, then $x < 4$ and $x \ge 0$. Therefore, $h(1) = 1^2 - 3 = -2$ and $h(1) = 2(1) + 1 = 3$.

This violates the requirement all functions must satisfy—the input $x = 1$ (among others) does not correspond to *one* output—so $h(x)$ is not a function.

Example 5: Consider the piecewise-defined function $m(x)$ as defined below.

$$m(x) = \begin{cases} 4x - x^2, x \le -1 \\ \sqrt{x+7}, -1 < x < 5 \\ x - 16, x \ge 5 \end{cases}$$

Evaluate the following:

(a) $m(-3)$

Solution: Look at the input restrictions to determine what you should plug $x = -3$ into. Because $-3 \le -1$, you should substitute it into the top expression, $4x - x^2$.

$$m(x) = 4x - x^2 \quad (\text{if } x \le -1)$$
$$m(-3) = 4(-3) - (-3)^2$$
$$= -12 - 9$$
$$= -21$$

(b) $m(9)$

Solution: Substitute $x = 9$ into the expression $x - 16$, because $9 \ge 5$.

$$m(x) = x - 16 \quad (\text{if } x \ge 5)$$
$$m(9) = 9 - 16$$
$$= -7$$

(c) $m(2)$

Solution: Because $x = 2$ falls between the numbers -1 and 5, you plug it into the middle expression, $\sqrt{x+7}$.

$$m(x) = \sqrt{x+7} \quad (\text{if } -1 < x < 5)$$
$$m(2) = \sqrt{2+7}$$
$$= \sqrt{9}$$
$$= 3$$

You've Got Problems

Problem 5: Given the function $f(x)$ as defined below, rank the values $f(-1)$, $f(0)$, and $f(1)$ in order from least to greatest.

$$f(x) = \begin{cases} 3x + 4, & x \le 0 \\ x - x^2, & x > 0 \end{cases}$$

The Least You Need to Know

◆ A relation pairs together inputs and outputs.

◆ A function is a relation whose inputs each result in a single corresponding output.

◆ When you plug one function into another, you perform composition of functions.

◆ A function and its inverse, when composed together, cancel each other out.

◆ A piecewise-defined function is made up of a handful of other functions saddled with input restrictions.

Chapter 16

Graphing Functions

In This Chapter

- Sketching graphs using brute force and finesse
- Characterizing functions based on their graphs
- Stretching, squishing, and moving graphs
- Determining the domain and range of functions

Even though you have factored and solved nonlinear equations for quite some time now, I haven't shown you their graphs yet. There's a reason for this: it may freak you out just a little bit. You're probably used to nice straight lines, and now, all of a sudden, curves get thrown into the mix. It's a world-shaking change.

It's sort of like sitting across from your mom at the breakfast table one morning, when she mentions in an off-handed manner that she's been a crime-fighting superhero named The Silver Heron for 12 years now. Your mom! This is the same kindly lady who brought cupcakes to your fourth-grade class when it was your birthday and tied empty bread bags around your feet before you put your snow boots on so that your socks would stay dry.

Of course, as cool as it would be to have a mom that could single-handedly thwart evildoers' plans to rule the world, it would take some getting used to. Metaphorically speaking, you're about to make the same sort of jump in graphing. While some minor things will stay the same (such as the coordinate system), you're going to have to learn some new techniques to conquer the new graphical foes. After all, Mom needs a sidekick.

Second Verse, Same as the First

The easiest way to graph a linear equation is to create a table, plugging in a few values for x to get the corresponding y-values. Although this chapter deals with nonlinear functions (instead of the equations of lines), you can use the same techniques to graph them. Plug a bunch of x-values into the function to get the corresponding $f(x)$ outputs. Then, plot the $(x, f(x))$ points on the coordinate plane.

Kelley's Cautions

Sometimes when you plug in values for x, you won't get a valid, real number output. That's okay. It just means that the number you plugged in is not a valid input for the function. (To use terminology from later in the chapter, that x-value is not in the "domain.") Keep plugging in different x's until you find some that work.

Critical Point

The u-shaped graph of a quadratic is called a parabola.

For instance, if you plug $x = 1$ into a function $f(x)$ to get $f(1)$, that corresponds to the point $(1, f(1))$ in the plane. After you plug in enough x-values, you'll begin to figure out what the graph looks like.

Example 1: Sketch the graph of $f(x) = (x + 2)^2 - 4$ using a table of values.

Solution: Because you're trespassing in the land of nonlinear graphs, you'll have to plot more than two or three points. This takes some experimenting. If you plug in x-values that give really large outputs, they're kind of hard to graph. Try other inputs, until you get some smaller results.

In the case of $f(x) = (x + 2)^2 - 4$, x-values less than -5 or greater than 1 produce unreasonably large results. However, if you evaluate $f(x)$ for integer values between -5 and 1, the graph of $f(x)$ begins to materialize.

x	$f(x) = (x+2)^2 - 4$	$f(x)$
-5	$f(-5) = (-5+2)^2 - 4 = (-3)^2 - 4$	5
-4	$f(-4) = (-4+2)^2 - 4 = (-2)^2 - 4$	0
-3	$f(-3) = (-3+2)^2 - 4 = (-1)^2 - 4$	-3
-2	$f(-2) = (-2+2)^2 - 4 = 0 - 4$	-4
-1	$f(-1) = (-1+2)^2 - 4 = (1)^2 - 4$	-3
0	$f(0) = (0+2)^2 - 4 = (2)^2 - 4$	0
1	$f(1) = (1+2)^2 - 4 = (3)^2 - 4$	5

To graph $f(x) = (x + 2)^2 - 4$, plot the points generated by the table: $(-5,5)$, $(-4,0)$, $(-3, -3)$, $(-2, -4)$, $(-1, -3)$, $(0,0)$, and $(1,5)$, as illustrated in Figure 16.1.

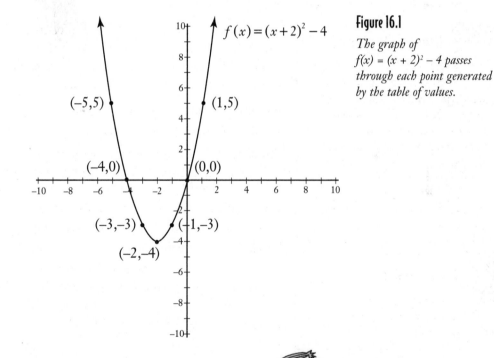

Figure 16.1

The graph of $f(x) = (x + 2)^2 - 4$ passes through each point generated by the table of values.

There's one thing I want to draw your attention to in Figure 16.1. This graph (unlike linear graphs) has two x-intercepts: $x = -4$ and $x = 0$. The x-intercepts are also called the *zeroes* of the function (because

Talk the Talk

The **zeroes** of a function $f(x)$ are its x-intercepts, the x-values that make the equation $f(x) = 0$ true.

$f(x)$ equals 0 for those x-values). In other words, the zeroes of $f(x)$ and the roots of the equation $f(x) = 0$ are exactly the same.

You've Got Problems

Problem 1: Use a table of values to graph $g(x) = -|x - 5| + 3$.

Two Important Line Tests

Predatory cats sleep a large portion of the day, so that's not a sign of weakness. Instead, you should ensure the large animal maintains its sharp eyes and keen reflexes; a loss in appetite or a general malaise toward food is often a symptom of a larger problem.

Wait a minute, I read the title of this section wrong. Those are important *lion* tests, not *line* tests. My mistake. There are two major line tests in algebra, one involving vertical lines and one that uses horizontal lines.

The Vertical Line Test

The *vertical line test* allows you to look at a graph and immediately know whether or not it's the graph of a function. To apply the test, ask yourself this question: "If I were to draw a vertical line *anywhere* on the graph, how many times would that line intersect the graph?" If your answer is, "The line hits the graph *no more than once*," you've got a function on your hands.

Talk the Talk

The **vertical line test** can determine whether or not a relation is a function by examining its graph. The **horizontal line test** uses a similar technique to determine whether or not a function is one-to-one.

Take a look at the three graphs in Figure 16.2, where vertical lines drawn on graphs (a) and (b) intersect the graph in two places. The lines are dotted to remind you that they're not actually part of the graph—they're just little strips used only for the test, like litmus paper that's thrown away after the experiment is over.

That's not true for graph (c), though. No vertical lines intersect that graph more than once, so the graph represents a function.

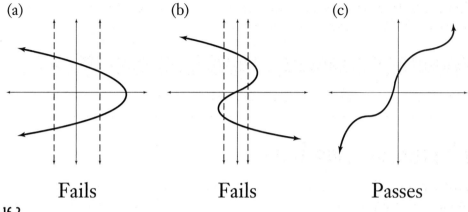

Figure 16.2

Graphs (a) and (b) fail the vertical line test, because you can draw vertical lines that intersect the graphs more than once.

How'd They Do That?

Vertical lines in the coordinate plane have equations that look like $x = c$, where c is a real number. When you draw vertical lines on a graph and find intersection points, you're identifying the outputs when $x = c$ is plugged into the function. Therefore, when a vertical line intersects a graph twice, there are two y-values (or outputs) that correspond to that x-value, which means the graph cannot represent a function.

The Horizontal Line Test

Once you know that a relation is a function, you can take it one step further and use the *horizontal line test* to determine whether or not that function is one-to-one. Remember, only one-to-one functions have inverses, because every output is paired with only one input. Don't bother trying to calculate the inverse of a function unless it first passes the horizontal line test.

The horizontal line test works a lot like the vertical line test did. Imagine a series of horizontal lines running across the graph. If none of those lines intersect the graph more than once, then the graph passes the test and is classified one-to-one. Vertical and horizontal lines in these tests are allowed to miss the graph altogether—they just can't hit it multiple times.

How'd They Do That?

Horizontal lines represent *y*-values (outputs), so if they intersect the graph more than once, that means multiple *x*'s have the same output and the function's not one-to-one.

Figure 16.3 shows the horizontal line test in action. Like Figure 16.2, many lines will intersect graphs (a) and (b) twice (including the dotted lines shown), so while all three graphs represent functions, only (c) is one-to-one.

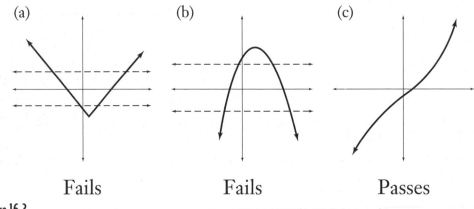

(a)   (b)   (c)

Fails   Fails   Passes

Figure 16.3

Only graph (c) passes the horizontal line test.

You've Got Problems

Problem 2: In Problem 1, you generated the graph of $g(x) = -|x - 5| + 3$. Verify that $g(x)$ is a function and determine whether or not it is one-to-one.

Determining Domain and Range

It's time to abandon the words "input" and "output" and embrace Official Math Terminology (drum roll, please). All of the numbers you can plug into a function collectively make up the *domain* of that function. Consider the function $f(x) = \dfrac{1}{x-2}$. There's one real number you can't substitute into $f(x)$: $x = 2$. Watch what happens when you try.

$$f(2) = \frac{1}{2-2} = \frac{1}{0}$$

That result does not compute, because dividing by 0 breaks the rules. (Other rules include: never get fractions wet and never feed them after midnight.) Because $x = 2$ is not a valid input, it is the only real number not in the domain of $f(x)$.

Once you figure out the domain of a function, plugging in all of those inputs gives you a collection of values called the *range*, the set of valid function outputs. However, the domain of a function is usually infinite, so physically plugging every possible input into the function is a bad idea.

Talk the Talk

The **domain** of a function is the set of valid inputs, and the **range** is the collection of outputs.

The best way to identify a function's domain and range is to perform modified versions of the vertical and horizontal line tests. To find the domain, ask yourself this question: "If I draw vertical lines all over graph, which ones will intersect the graph, and which ones won't?" All of the vertical lines that hit the graph represent x-values in the domain. Similarly, imagine lots of horizontal lines superimposed over the function. Any such lines that intersect the graph represent y-values in the range.

Example 2: Determine the domain and range of the function $f(x) = \sqrt{x+3} - 1$, given its graph in Figure 16.4.

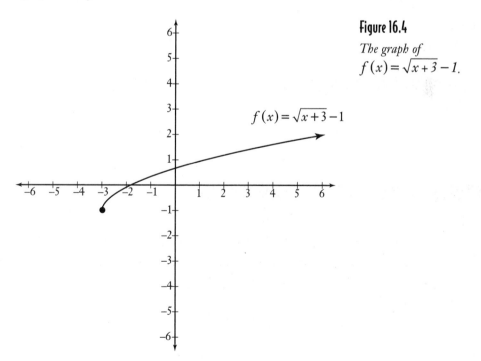

Figure 16.4

The graph of
$f(x) = \sqrt{x+3} - 1.$

$f(x) = \sqrt{x+3} - 1$

Solution: The graph extends infinitely to the right but stops abruptly at the point (–3,–1). Any vertical lines left of $x = -3$ (like $x = -3.5$, $x = -4$, $x = -5$, and so on) will not intersect the graph, whereas vertical lines for x-values greater than or equal to $x = -3$ will. Therefore the domain of $f(x)$ is $x \geq -3$

Critical Point

Creating a zero denominator and causing a negative value within a radical are the two most common reasons a number must be excluded from the domain of a function.

And in case you're wondering where in the world the graph of $f(x)$ came from in Example 2, don't worry—you'll learn how to graph that and many other functions by the end of the chapter.

Now imagine horizontal lines splattered all over the coordinate plane. The horizontal line $y = -1$ hits the point (–3,–1), but any horizontal lines below that will pass harmlessly beneath the graph. On the other hand, any horizontal line above $y = -1$ will hit the graph somewhere. Even though the graph in Figure 16.4 is shallow, it will eventually reach infinite heights. Therefore, the range of the function is $y \geq -1$.

You've Got Problems

Problem 3: Identify the domain and range of $g(x) = -|x - 5| + 3$, the function referenced in Problems 1 and 2 earlier in this chapter.

Important Function Graphs

It's not nice of me to keep talking about things you can do with a function graph without actually helping you make the graph, too. Of course, you can make any graph in the world by creating a table of values, but there are quicker ways to sketch graphs without investing years of your life. However, before you can create those sketches, you first need to memorize some graphs.

In Figure 16.5, you'll find the five basic graphs you'll make in algebra. Memorize their general shapes and important characteristics—you're making quick sketches, not exact graphs. If you want to understand how those graphs were created, create a table of values and plot points (as explained earlier, in Example 1). However, these graphs are the building blocks for the sketching method explained in the next section, so it's important that you can create them from memory rather than generating them from scratch every time.

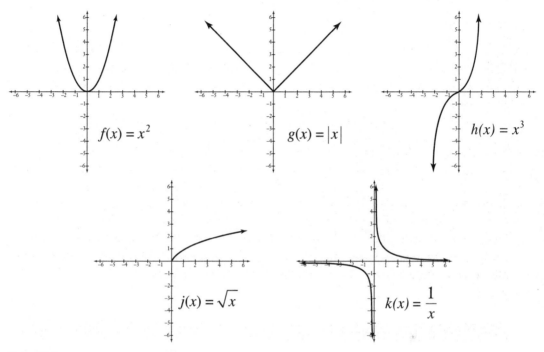

$f(x) = x^2$

$g(x) = |x|$

$h(x) = x^3$

$j(x) = \sqrt{x}$

$k(x) = \dfrac{1}{x}$

Figure 16.5

The five fundamental algebraic graphs you should memorize.

Here are brief descriptions of each graph to help you store them away in the recesses of your brain. (By the way, I used different letters for different functions so you can tell them apart—like $f(x)$, $g(x)$, and $h(x)$—but those letters are arbitrary, so don't pay them much attention.)

- $f(x) = x^2$ **(Domain: all reals; Range: $y \geq 0$)**

    You can square any real number and the result will always be positive. Notice that the graph changes direction at the origin. As you trace it from left to right, it decreases until (0,0) and then increases after that.

Critical Point

Figure 16.5 contains graphs of degree two (x^2) and degree three (x^3), but not a linear equation with degree 1. That's because linear equations and graphs are covered in Chapters 5 and 6—just treat $f(x)$ like the variable y.

◆ $g(x) = |x|$ **(Domain: all reals; Range: $y \geq 0$)**

The absolute value function accepts any real number input but only lets positive numbers out. It, like $f(x) = x^2$, changes direction at the origin.

◆ $h(x) = x^3$ **(Domain: all reals; Range: all reals)**

The cubic function produces negative outputs when fed negative inputs. The graph's key feature is a little twist near $x = 0$ that forces it to pass through the origin.

◆ $j(x) = \sqrt{x}$ **(Domain: $x \geq 0$; Range: $y \geq 0$)**

You can take the square root of real numbers only when they're positive. In addition, a square root always outputs a positive value, so the domain and range are harshly restricted here. In fact, the graph only appears in the first quadrant, because in each of the other quadrants either x or y (or both x and y) are negative.

◆ $k(x) = \dfrac{1}{x}$ **(Domain: $x \neq 0$; Range: $y \neq 0$)**

Even though the graph gets close to the vertical line $x = 0$ and the horizontal line $y = 0$, it won't touch either. Plugging 0 into $\dfrac{1}{x}$ breaks that "don't divide by zero" rule, so $x = 0$ is not in the domain. Furthermore, there's nothing you can plug into $\dfrac{1}{x}$ that produces an output of 0, so 0 is not in the range, either. This is the only graph of the five that does not contain the origin for those reasons. However, every real number *except* 0 belongs in both the domain and range.

Critical Point

The graph of $k(x) = \dfrac{1}{x}$ gets close to, but never touches, the lines $x = 0$ and $y = 0$. Those lines are the asymptotes of the graph.

Spend some quality time with these graphs to memorize them. Make sure that, given only an equation like $j(x) = \sqrt{x}$, you could draw the graph without peeking at Figure 16.5. After you can do that, it's time to put all that boring memorization to good use.

Graphing Function Transformations

Most of the graphs you create in algebra are just funky versions of the graphs in Figure 16.5. Sketching the vast majority of graphs is as easy as transforming one of those five basic building blocks. What does it mean to "transform" the graphs? Transformations are simple adjustments like moving them around the coordinate plane, stretching or squishing them a little, or flipping them around one of the axes.

All these things can be accomplished by sticking a number here or a negative sign there—small changes make a big difference in a graph.

Keep in mind that the sketches you create using function transformations are only as exact as the graph you start with, but don't break your back trying to be perfect. If you needed a perfect graph, you could always plot 500 points or use a graphing calculator. In most cases, quick sketches are fine, and besides, learning to graph this way teaches you how all the coefficients and constants in a function affect that function's graph.

Making Functions Flip Out

Like a lion tamer at the circus, who need only blow his whistle to make hungry lions do an incredibly complicated trick (like waltzing with each other to the tune of "Do That to Me One More Time" by Captain and Tennille), very little effort is required to reflect a graph (or "flip it over") an axis—just stick a negative sign in the right place. Here's how the graph of a function $f(x)$ is affected when you start inserting negative signs:

- **$-f(x)$ is the reflection across the x-axis.** In other words, if you multiply the function itself by -1, the graph will flip over the x-axis. Each of its x-coordinates will stay the same, but every y-coordinate will be the opposite. Therefore, if $f(x)$ contains the point (a,b), $-f(x)$ contains $(a,-b)$.

- **$f(-x)$ is the reflection across the y-axis.** If you plug $-x$ into the function instead of x, the x-coordinate changes to its opposite.

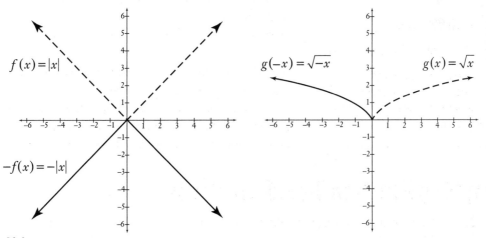

Figure 16.6

The location of the negative sign dictates what sort of reflection will occur.

Each of these transformations is illustrated by Figure 16.6. The left-hand graph shows that $f(x) = |x|$ and $-f(x) = -|x|$ are reflections of each other across the x-axis. In the right-hand graph, $g(x) = \sqrt{x}$ is reflected across the y-axis when you plug $-x$ into the function.

Stretching Functions

You can use more than negative signs to transform functions; numbers have a profound effect as well. Again, the placement of the number determines whether it will affect either the function's *height* or its *width*:

◆ **The graph of $a \cdot f(x)$ is a times as tall as $f(x)$.** If you multiply a function by a number, then each of the outputs get multiplied by that number. For instance, each output of some function $4g(x)$ is exactly four times as high as the corresponding output of $g(x)$. If a is between 0 and 1, then the graph gets squished toward the x-axis instead of stretched vertically. For instance, each point on the graph of $\frac{1}{2}f(x)$ is half the distance from the x-axis than the corresponding point on $f(x)$.

◆ **The graph of $g(bx)$ is $\frac{1}{b}$ times as wide as $g(x)$.** This rule is a little bizarre.

Think of it this way: If you're plugging bx into a function instead of x (where b is a real number), then the graph will get horizontally squished toward the origin by a factor of b if $b > 1$. If $b < 1$, the graph gets horizontally stretched out by a factor of b.

Critical Point

Remember, $\frac{1}{b}$ means "the reciprocal of b," so $\frac{1}{1/3}$ equals 3, because the reciprocal of $\frac{1}{3}$ is 3.

As you can see in the left-hand graph of Figure 16.7, multiplying $f(x) = \sqrt{x}$ by 2 causes its graph to stretch twice its original height. The right-hand graph shows that plugging $\frac{1}{3}x$ into $g(x) = x^2$ causes the graph to stretch out $\frac{1}{1/3} = 3$ times as wide.

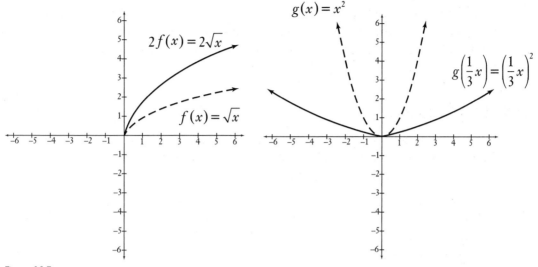

Figure 16.7

Multiplying a function by a number causes the graph to stretch away from or get squished toward one of the axes.

Moving Functions Around

The final transformation also involves numbers, but this time, instead of multiplying them you'll add them to the function.

- **Adding to or subtracting from a function moves its graph up or down, respectively.** That means the graph of $f(x) + 7$ is the graph of $f(x)$ moved up seven units, and the graph of $g(x) - 1$ is the graph of $g(x)$ moved down one unit.

- **Plugging $x + a$ into a function moves its graph a units left or right.** If $a < 0$, the graph will move right, and if $a > 0$, the graph moves left a units. This is the opposite of what intuition tells you: subtracting a causes the graph to move *right*, and adding a causes the graph to move *left*.

Therefore, adding to or subtracting from a function moves it vertically, but doing the same thing *inside* a function (adding to or subtracting from the x-input) moves it horizontally. Figure 16.8 illustrates these horizontal and vertical graph shifts.

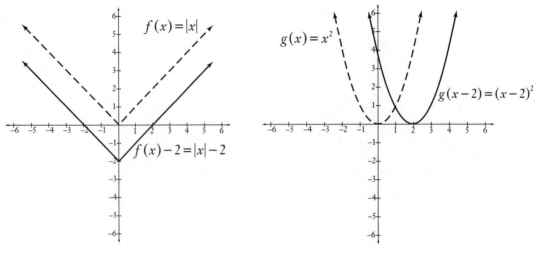

Figure 16.8

Adding a constant to the input of a function moves the graph of the function horizontally, and adding a constant to the output moves the graph vertically.

Multiple Transformations

When you need to graph a function that has more than one transformation applied to it, you should perform the transformations in the order this chapter presented them:

1. Reflections

2. Stretching/squishing

3. Vertical/horizontal shifting

It's the same order I follow when I wake up in the morning: look in the mirror (reflect), do a few sit-ups (stretch), and then go to work (to complete my shift).

Example 3: Graph $f(x) = -2|x+3| + 5$.

Solution: Start with a graph of the basic absolute value function $|x|$. Notice that the absolute value in $f(x)$ is multiplied by -2, so it is flipped across the x-axis and stretched vertically by a factor of 2. Finally, the graph is moved 3 units left and 5 units up. After the smoke has cleared, you end up with the graph in Figure 16.9.

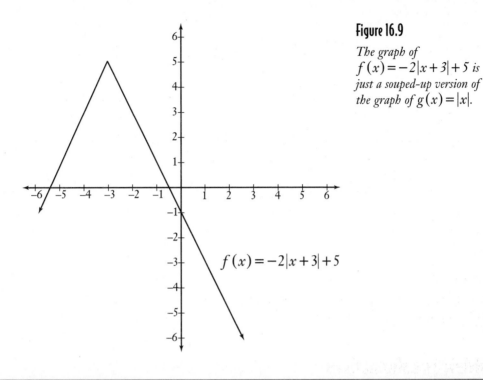

$$f(x) = -2|x+3|+5$$

You've Got Problems

Problem 4: Graph $g(x) = (-x)^3 - 2$.

The Least You Need to Know

♦ The domain of a function is the collection of all its valid inputs, and the range of a function consists of all the valid outputs.

♦ The vertical line test determines whether or not a graph represents a function, and the horizontal line test determines whether or not the function is one-to-one.

♦ Inserting a negative into a function causes its graph to reflect about one of the axes.

♦ Multiplying a function by a number will cause its graph to stretch or squish horizontally or vertically.

♦ Adding to or subtracting from a function causes either a horizontal or vertical shift in its graph.

Part

Please, Be Rational!

Way, way back in Chapter 2, fractions roamed the earth. With them came concepts like the least common denominator (along with other terms that make even the stoutest and bravest of humankind cry to this day) and operations on those fractions, such as addition and division. You might have thought that fractions were ancient history, but in this part, even fiercer fractions rear their heads—fractions with polynomials in their numerators and denominators.

Chapter 17

Rational Expressions

In This Chapter

- ◆ Reducing fractions containing polynomials
- ◆ Performing basic operations on rational expressions
- ◆ Eliminating fractions within fractions

There are lots of fancy words in the English language that describe nasty things; even the most distressing concepts don't sound so bad if they have a pretty name. For example, who wants to go to "war" when a "conflict" sounds so much less threatening? Most people are opposed to "tax increases" but wouldn't be automatically opposed to an "income adjustment."

In mathematics, when you mention the word "fraction," people panic. They sweat, their eyes dart about nervously, and then they sprint from the room. As a former high school math teacher, I know this from experience.

One morning, in a basic algebra class, I announced that we would be discussing fractions, and a student in the front row bolted like a frightened deer. He ran for the door, but in the process managed to snag his pant leg on another student's desk. Caught in his instinctual flight response, he couldn't think straight to turn around and free his clothing from the obstruction and, instead, *chewed off his own leg to get free.*

Well, maybe that's a bit of an exaggeration, but the innate fear of fractions that's shared by many people is no less real. (He actually tripped over a desk on the way to sharpen his pencil, but that's not as dramatic a story.) So, in a thinly veiled attempt to keep things calm, this chapter is titled "Rational Expressions" instead of "Fractions are Back, and This Time It's Personal."

Chapter 1 explained that a rational number (and likewise a rational expression) is a number that can be expressed as a fraction, but for some reason the word "rational" doesn't have the same pain and fear attached to it as the word "fraction"—just like the word "child entertainer" doesn't inspire the fear that the phrase "scary maniacal clown" does.

Simplifying Rational Expressions

The plan of action for this chapter should be familiar. The introduction of a new (or, in this case, old but suddenly complicated) concept begins by learning how to add, subtract, multiply, and divide the new things. However, rational expressions are fractions, so you need to add one more skill to the list—reducing fractions.

Before you simplify fractions containing polynomials, think about the process used to simplify fractions in Chapter 2. Back then, I told you that the best way to reduce a fraction was to divide both its numerator and denominator by any common factors. Now, I want to show you why that works.

Critical Point

If you started with $2 \cdot 6$ in your attempt to determine the prime factorization of 12, you got the same final result: $2 \cdot 6 = 2 \cdot (3 \cdot 2) = 2 \cdot 2 \cdot 3$. That's because every number has a unique prime factorization (according to something called the Fundamental Theorem of Algebra).

Consider the fraction $\frac{12}{30}$. Rewrite the numerator and denominator as a product of prime factors. This means breaking down each number into a product and then factoring those numbers until all you're left with are prime numbers. For example, you could multiply $4 \cdot 3$ to get 12 (the numerator). However, that's not a prime factorization because 4 isn't a prime number, so rewrite 4 as a product as well: $4 \cdot 3 = 2 \cdot 2 \cdot 3$. All of those numbers are prime, so that's the prime factorization you're looking for.

Now it's time to factor the denominator, 30, into primes. Notice that $3 \cdot 10 = 30$, but 10 is not a prime number. Factor it as $2 \cdot 5$, and you get the prime factorization of 30: $2 \cdot 3 \cdot 5$. Rewrite the fraction $\dfrac{12}{30}$ using those strings of factors.

$$\frac{12}{30} = \frac{2 \cdot 2 \cdot 3}{2 \cdot 3 \cdot 5}$$

If any factor in the numerator has a twin in the denominator, cross out those two twin numbers. Here, you should cross out a pair of 2's and a pair of 3's.

$$\frac{\cancel{2} \cdot 2 \cdot \cancel{3}}{\cancel{2} \cdot \cancel{3} \cdot 5}$$

The reduced form of the fraction is $\dfrac{2}{5}$.

In a nutshell, to reduce (or simplify) a fraction, factor the numerator and denominator and then eliminate matching pairs of factors on opposite sides of the fraction bar. Pretty simple, eh? Factor, cross out, game over. Do the same thing when you find polynomials in the fraction.

Example 1: Simplify the expression $\dfrac{9-x^2}{x^2+x-12}$.

Solution: Begin by factoring the numerator and denominator. Notice that the numerator is the difference of perfect squares.

$$\frac{(3-x)(3+x)}{(x+4)(x-3)}$$

Uh oh—no matching factors. Even though $3 - x$ and $x - 3$ are close, they don't match. However, they will match if you factor -1 out of $3 - x$: $3 - x = -1(-3 + x)$.

$$\frac{-1(-3+x)(3+x)}{(x+4)(x-3)}$$

If you think about it, $-3 + x$ and $x - 3$ are equal (thanks to the commutative property for addition—you're just adding x and -3 in a different order each time), so you can cross out those matching factors.

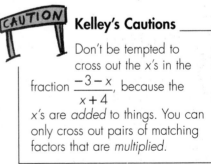

Kelley's Cautions

Don't be tempted to cross out the x's in the fraction $\dfrac{-3-x}{x+4}$, because the x's are *added* to things. You can only cross out pairs of matching factors that are *multiplied*.

$$\frac{-1(\cancel{-3+x})(3+x)}{(x+4)(\cancel{x-3})} = \frac{-1(3+x)}{x+4}$$

It's common to leave a reduced rational expression in factored form like this (especially later, when you'll have lots of factors left in the numerator and denominator), but it's just as correct to distribute the −1 in the numerator, if you are so inclined.

$$\frac{-3-x}{x+4}$$

You've Got Problems

Problem 1: Simplify the expression $\dfrac{2x^3 + 5x^2 - 3x}{6x^2 + 7x - 5}$.

Combining Rational Expressions

Once the fancy vocabulary has been stripped away, rational expressions are just fractions, so adding or subtracting them will still require you to find a least common denominator (LCD). Unfortunately, the Bubba technique described in Chapter 2 doesn't work once variables get tossed into the mix—you'll need to use a more formal method:

1. Factor all denominator polynomials completely.

2. Make a list that contains one copy of each factor, all multiplied together.

3. The power of each factor in that list should be the highest power that the factor is raised to in any denominator.

4. The list of factors and powers, multiplied together, is the LCD.

Critical Point

You can use the formal method for calculating the LCD with numbers as well—just use prime factorizations of the numbers. For instance, to calculate the LCD of $\dfrac{5}{36}$ and $\dfrac{11}{40}$, start by generating the prime factorizations of the denominators.

$$36 = 4 \cdot 9 = (2 \cdot 2) \cdot (3 \cdot 3) = 2^2 \cdot 3^2$$
$$40 = 5 \cdot 8 = 5 \cdot (2 \cdot 2 \cdot 2) = 2^3 \cdot 5$$

There are three different factors: 2, 3, and 5. The LCD will be the product of those factors, when each one is raised to the highest power it achieves in the factorizations. The factor 2 appears twice, as 2^2 and 2^3; you should use the 2^3 version in the LCD because it has the higher power. Numbers 3 and 5 only appear once, so use the powers attached to them: 3^2 and 5^1.

Therefore, the LCD of $\dfrac{5}{36}$ and $\dfrac{11}{40}$ is:

$$2^3 \cdot 3^2 \cdot 5 = 8 \cdot 9 \cdot 5 = 360$$

If finding an LCD is like going to a dance, then consider this method your dressy suit pants, whereas the Bubba technique would be a well-worn pair of jeans. Both keep you clothed, but neither is appropriate in all circumstances—it depends upon the kind of dance you're going to. There's nothing wrong with going informal if a fraction contains only numbers, but if you spot variables, you have to get all gussied up with the formal method. Actually, if given the choice, I'd prefer not going at all, because my dance technique looks remarkably like someone walking into an electrified fence.

Once you figure out the least common denominator, you can force all the denominators to match and then add the fractions together.

Example 2: Simplify the expression.

$$\frac{2x+4}{x^2-12x+36} - \frac{3x}{x^2+2x-48}$$

Solution: Begin by factoring the denominators.

$$\frac{2x+4}{(x-6)^2} - \frac{3x}{(x+8)(x-6)}$$

There are two different factors in the denominators, $x - 6$ and $x + 8$, so list them multiplied together: $(x - 6)(x + 8)$. Notice that $x - 6$ appears in both denominators, so to create the LCD, use the one raised to the higher power: $(x - 6)^2$ instead of $(x - 6)^1$. There–fore, the LCD is $(x - 6)^2(x + 8)$. Don't multiply the factors of the LCD together—leave them as is.

Compare each denominator with the LCD. What factor, if any, does each denominator need to match the LCD? The left fraction needs $x + 8$ and the right fraction needs another $x - 6$. Multiply both the numerator and denominator of each fraction by the factor or factors it needs to make its denominator complete.

$$\left(\frac{x+8}{x+8}\right)\cdot\frac{2x+4}{(x-6)^2} - \frac{3x}{(x+8)(x-6)}\cdot\left(\frac{x-6}{x-6}\right)$$

Multiply only the numerators together, leaving the (now-matching) denominators alone.

$$\frac{2x^2+20x+32}{(x-6)^2(x+8)} - \frac{3x^2-18x}{(x-6)^2(x+8)}$$

Now that the fractions have common denominators, you can combine their numerators and write them together over the common denominator. The second fraction is subtracted, so the negative sign will need to be distributed through its entire numerator.

$$\frac{2x^2 + 20x + 32 - (3x^2 - 18x)}{(x-6)^2(x+8)}$$

$$= \frac{2x^2 + 20x + 32 - 3x^2 + 18x}{(x-6)^2(x+8)}$$

$$= \frac{(2x^2 - 3x^2) + (20x + 18x) + 32}{(x-6)^2(x+8)}$$

$$= \frac{-x^2 + 38x + 32}{(x-6)^2(x+8)}$$

At this point, you should check to see if the numerator can be factored. If it can, then check for matching factors in the numerator and denominator to see if the fraction can be simplified; in this case, the numerator is prime. The answer $\frac{-x^2 + 38x + 32}{(x-6)^2(x+8)}$ is fine and dandy, but if you absolutely must multiply the terms in the denominator, you get $\frac{-x^2 + 38x + 32}{x^3 - 4x^2 - 60x + 288}$.

You've Got Problems

Problem 2: Simplify the expression $\frac{x}{x^2 - 4} + \frac{8}{x^2 - 7x + 10}$.

Multiplying and Dividing Rationally

My nine-month-old son Nicholas has been eating lumpy food for a couple of weeks now. He's not up to solid food yet, and can't actually feed himself, but if you spoon anything that's liquefied into his maw, he will slurp it down. Sometimes, that is. There's always the chance he'll sneeze with no warning and spray my face and glasses with half-chewed green bean casserole (which is a real treat if I happened to have my mouth open at the time).

He's also good at the tongue raspberry (which covers as effectively as a spray can) and the less-fancy move of simply spitting whatever's in his mouth onto his lap. Of course, his digestive system works the same as mine, and once he decides to eat the food in his mouth, the mechanics work the same way they do with me—chew, swallow, digest, create foul-smelling diaper. (Actually, the final step there is all his own.)

Suffice it to say that even though he and I digest the same way, he is a messier eater. (Unlike him, I rarely get food in my eyebrows anymore.) By extension, you use the same technique to multiply and divide fractions, whether they contain simple numbers or clunkier polynomials. Even though they work the same way, that doesn't mean it *feels* the same. Like my son, polynomial fractions tend to get a bit messy, as you probably noticed when adding and subtracting them in the previous section.

Remember, you don't need common denominators to calculate a product or a quotient—just multiply the numerators together and divide by the denominators multiplied together.

$$\frac{a}{b} \cdot \frac{c}{d} = \frac{ac}{bd}$$

Since I'm dredging up the past by reviewing fraction multiplication, here's one more historical tidbit. Division is the same as multiplying by a reciprocal; to find a quotient of two fractions, flip the fraction that you're dividing by upside down and change the division sign into a multiplication sign.

$$\frac{a}{b} \div \frac{c}{d} = \frac{a}{b} \cdot \frac{d}{c} = \frac{ad}{bc}$$

Example 3: Simplify the expressions.

(a) $\dfrac{2x-1}{x^2-2x-15} \cdot \dfrac{x-5}{2x^2-5x-3}$

Solution: Rewrite this expression as one fraction, keeping the numerators on top and the denominators on the bottom.

$$\frac{(2x-1)(x-5)}{\left(x^2-2x-15\right)\left(2x^2+5x-3\right)}$$

Instead of multiplying the denominator, factor it.

$$\frac{(2x-1)(x-5)}{(x-5)(x+3)(2x+1)(x+3)}$$

Kelley's Cautions _____

Don't forget to simplify your answers after you're finished multiplying or dividing.

If all of the factors in the numerator or the denominator get eliminated, place a 1 where the factors used to be (like in the numerator of the answer to Example 3a). Don't be tempted to put a 0 there, even though nothing seems to be left.

Simplify the fraction by crossing out pairs of matching factors in the numerator and denominator.

$$\frac{(2x-1)(x-5)}{(x-5)(x+3)(2x-1)(x+3)} = \frac{1}{(x+3)^2} \text{ or } \frac{1}{x^2+6x+9}$$

Either of those forms of the answer is correct.

(b) $\dfrac{x^4-4x^3-21x^2}{4x^2-9} \div \dfrac{2x^3-19x^2+35x}{8x^3+27}$

Solution: Start by taking the reciprocal of the right-hand fraction and changing this from a division to a multiplication problem. The left-hand fraction remains unchanged.

$$\frac{x^4-4x^3-21x^2}{4x^2-9} \cdot \frac{8x^3+27}{2x^3-19x^2+35x}$$

Write the product as a single fraction, factoring each of the four expressions. (If you can't remember the formula to factor the sum of perfect cubes, flip back to Chapter 11.)

$$\frac{(x^4-4x^3-21x^2)(8x^3+27)}{(4x^2-9)(2x^3-19x^2+35x)}$$

$$= \frac{x^2(x^2-4x-21)(8x^3+27)}{(4x^2-9)(x)(2x^2-19x+35)}$$

$$= \frac{x^2(x-7)(x+3)(2x+3)(4x^2-6x+9)}{(2x+3)(2x-3)(x)(x-7)(2x-5)}$$

To make simplifying easier, rewrite the monomial factor x^2 in its factored form, $x \cdot x$. When you reduce the fraction, one of those x's will cancel out with the x in the denominator.

$$\frac{x \cdot x (x-7)(x+3)(2x+3)(4x^2-6x+9)}{(2x+3)(2x-3)(x)(x-7)(2x-5)}$$

$$= \frac{x(x+3)(4x^2-6x+9)}{(2x-3)(2x-5)}$$

Once again, there's no need to multiply those factors together—a final answer in factored form is just dandy. However, if you feel compelled to multiply, you'll end up with $\dfrac{4x^4+6x^3-9x^2+27x}{4x^2-16x+15}$.

You've Got Problems

Problem 3: Simplify the expression $\dfrac{x^2-4x-12}{3x^2-10x-8} \div \dfrac{x^2-3x-18}{3x+2}$.

How'd They Do That?

One of the exponential rules from Chapter 3 stated that $\frac{x^a}{x^b} = x^{a-b}$. Example 3(b) demon-

strates why this is true. In that example, you rewrote x^2 as $x \cdot x$ and then canceled out factors to reduce the fraction.

Consider the fraction $\frac{x^{10}}{x^7}$; according to the above rule, $\frac{x^{10}}{x^7} = x^{10-7} = x^3$. If you write x^{10} and x^7 as strings of x's and cancel numerator/denominator pairs of them out, you get the same answer. Seven of the x's in the numerator cancel with the seven x's in the denominator, leaving behind $\frac{x^3}{1} = x^3$:

$$\frac{x^{10}}{x^7} = \frac{\cancel{x} \cdot \cancel{x} \cdot \cancel{x} \cdot \cancel{x} \cdot \cancel{x} \cdot \cancel{x} \cdot \cancel{x} \cdot x \cdot x \cdot x}{\cancel{x} \cdot \cancel{x} \cdot \cancel{x} \cdot \cancel{x} \cdot \cancel{x} \cdot \cancel{x} \cdot \cancel{x}} = \frac{x \cdot x \cdot x}{1} = x^3$$

Encountering Complex Fractions

If you hate fractions, then you'll be no fan of *complex fractions*. Just the name alone sounds scary, right? Fractions are hard enough, but *complex* fractions? Great! I imagine that brain surgery is pretty hard to do, but *complex* brain surgery sounds even worse. Actually, your gut fear is probably unjustified because the term "complex fraction" is false advertising for two reasons:

◆ The word "complex" suggests that the fractions contain complex numbers, but they don't.

◆ You already know how to work with complex fractions. You just don't *know* that you know how to. (But you know now.)

Enough mystery—let's cut to the chase. A complex fraction is a fraction that contains a fraction in its numerator or denominator (or both). Leaving final answers as complex fractions is bad form because it looks unnecessarily complicated. Luckily, fixing them is pretty simple—just translate them into division problems and divide like you did in Example 3.

Talk the Talk

A complex fraction (or compound fraction) contains fractions in its numerator or denominator (or both). It's sort of a double-decker fraction.

Example 4: Simplify the complex fraction below.

$$\frac{\dfrac{3x}{x-2}}{\dfrac{9x^2}{7x-14}}$$

Solution: Rewrite the complex fraction as a quotient—the top fraction divided by the bottom fraction.

$$\frac{3x}{x-2} \div \frac{9x^2}{7x-14}$$

Suddenly, this looks a lot like Example 3(b). A little multiplication by the reciprocal should do the trick. Don't forget to factor.

$$\frac{3x}{x-2} \cdot \frac{7x-14}{9x^2}$$
$$= \frac{3x}{x-2} \cdot \frac{7(x-2)}{9 \cdot x \cdot x}$$
$$= \frac{3 \cdot x \cdot 7 \cdot (x-2)}{(x-2) \cdot 9 \cdot x \cdot x}$$

Simplify the fraction.

$$\frac{3 \cdot \cancel{x} \cdot 7 \cdot \cancel{(x-2)}}{\cancel{(x-2)} \cdot 9 \cdot \cancel{x} \cdot x}$$
$$= \frac{3 \cdot 7}{9 \cdot x}$$
$$= \frac{21}{9x}$$

You're not quite done yet, because you can simplify the fraction further: 21 and 9 are both divisible by 3.

$$\frac{7}{3x}$$

You've Got Problems

Problem 4: Simplify the complex fraction.

$$\frac{\dfrac{x^2-7x+12}{x+2}}{\dfrac{x-3}{x^2+4x+4}}$$

The Least You Need to Know

- To simplify rational expressions, factor the numerator and denominator and cancel out pairs of factors that appear in each.

- You can only add or subtract rational expressions that have common denominators.

- Change rational division problems into multiplication problems by taking the reciprocal of the fraction you're dividing by.

- Complex fractions are really just quotients in disguise.

Rational Equations and Inequalities

In This Chapter

◆ Solving equations containing fractions

◆ Simplifying proportions using cross multiplication

◆ Exploring direct and indirect variation

◆ Solving rational inequalities

Knowing *how* to add and multiply fractions is important, but knowing *when* to apply those skills is equally important. In this chapter, fractions start popping up all over the place and invade the concepts discussed in the preceding chapters. What do you do when an equation contains fractions? How in the world do you solve an inequality featuring fractions? How are you supposed to react if a fraction sneaks into your house when you're asleep and tries to steal that big jar of pennies you're not sure what to do with?

Solving Rational Equations

In the world of math, when you don't like something, you can manipulate the rules of the universe and make things disappear. Case in point: most mathematicians dislike fractions. It's not because they don't understand them, it's just that the constant need of common denominators is annoying, so they'd prefer to eliminate them whenever possible.

The easiest way to make the fractions in an equation disappear is to multiply everything by the least common denominator (LCD) of the fractions.

Kelley's Cautions

Multiplying an equation by something containing an x occasionally introduces a few false solutions, so plug all of your answers back into the original equation to make sure they're valid.

Example 1: Solve the equation.

$$\frac{1}{x+5} + \frac{x}{x+2} = \frac{2x-1}{x^2+7x+10}$$

Solution: This equation contains three rational expressions; your first goal will be to eliminate all of those fractions to make solving the equation easier. Start by factoring every expression you can—in this case, $x^2 + 7x + 10$ is factorable.

$$\frac{1}{x+5} + \frac{x}{x+2} = \frac{2x-1}{(x+5)(x+2)}$$

Multiply both sides of the equation by the least common denominator $(x + 5)(x + 2)$ to eliminate the fractions. Technically speaking, you'll multiply the equation by $\frac{(x+5)(x+2)}{1}$. The 1 in the denominator reminds you to multiply each numerator by the LCD and leave the denominators unchanged.

$$\frac{(x+5)(x+2)}{1} \cdot \left[\frac{1}{x+5} + \frac{x}{x+2}\right] = \frac{(x+5)(x+2)}{1} \cdot \left[\frac{2x-1}{(x+5)(x+2)}\right]$$

$$\frac{(x+5)(x+2)}{x+5} + \frac{x(x+5)(x+2)}{x+2} = \frac{(x+5)(x+2)(2x-1)}{(x+5)(x+2)}$$

Simplify the fractions.

$$\frac{\cancel{(x+5)}(x+2)}{\cancel{x+5}} + \frac{x(x+5)\cancel{(x+2)}}{\cancel{x+2}} = \frac{\cancel{(x+5)}\cancel{(x+2)}(2x-1)}{\cancel{(x+5)}\cancel{(x+2)}}$$

All of the denominators are equal to 1, but there's no need to write a denominator of 1.

$$(x+2)+x(x+5)=(2x-1)$$
$$x+2+x^2+5x=2x-1$$
$$x^2+6x+2=2x-1$$

Set the equation equal to zero by subtracting $2x$ from, and adding 1 to, both sides.

$$x^2+6x-2x+2+1=0$$
$$x^2+4x+3=0$$

Solve the quadratic equation by factoring.

$$(x+3)(x+1)=0$$

$$x+3=0 \qquad \text{or} \qquad x+1=0$$
$$x=-3 \qquad\qquad x=-1$$

Plug both of those solutions into the original equation. Because you get true statements, they are valid answers.

$\boxed{\text{Check } x=-3}$	$\boxed{\text{Check } x=-1}$
$\dfrac{1}{x+5}+\dfrac{x}{x+2}=\dfrac{2x-1}{x^2+7x+10}$	$\dfrac{1}{x+5}+\dfrac{x}{x+2}=\dfrac{2x-1}{x^2+7x+10}$
$\dfrac{1}{-3+5}+\dfrac{-3}{-3+2}=\dfrac{2(-3)-1}{(-3)^2+7(-3)+10}$	$\dfrac{1}{-1+5}+\dfrac{-1}{-1+2}=\dfrac{2(-1)-1}{(-1)^2+7(-1)+10}$
$\dfrac{1}{2}+\dfrac{-3}{-1}=\dfrac{-6-1}{9-21+10}$	$\dfrac{1}{4}+\dfrac{-1}{1}=\dfrac{-2-1}{1-7+10}$
$\dfrac{1}{2}+3=\dfrac{-7}{-2}$	$\dfrac{1}{4}-1=\dfrac{-3}{4}$
$\dfrac{1}{2}+\dfrac{6}{2}=\dfrac{7}{2}$ $\boxed{\text{True}}$	$\dfrac{1}{4}-\dfrac{4}{4}=\dfrac{3}{4}$ $\boxed{\text{True}}$

You've Got Problems

Problem 1: Solve the equation $\dfrac{x+3}{x-8}+\dfrac{x}{x^2-6x-16}=1$.

Proportions and Cross Multiplying

Of all the rational equations you'll see as an algebra student, one of the most common will be the *proportion*, an equation in which two fractions are set equal to one another, like this: $\dfrac{a}{b}=\dfrac{c}{d}$.

Talk the Talk

A **proportion** consists of two fractions set equal to each other. Proportions are often solved using **cross multiplication,** multiplying the numerator of one fraction by the denominator of the other and setting those products equal.

You can solve a proportion (just like any other rational equation) if you multiply it by the least common denominator of the fractions and simplify. However, there's a nifty shortcut called *cross multiplication* that works exclusively for proportions: multiply each numerator by the denominator of the other fraction and set the results equal, as illustrated by Figure 18.1.

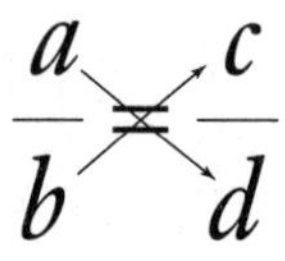

Figure 18.1

Cross multiplying eliminates the denominators of a proportion and doesn't require you to calculate the least common denominator.

$$\frac{a}{b} \diagdown \frac{c}{d}$$

$$a \cdot d = b \cdot c$$

Example 2: Solve the equation $\dfrac{x+3}{x-1} = \dfrac{x+6}{2}$.

Solution: Because the equation is a proportion, cross multiply to eliminate the fractions.

$$(x+3)\cdot 2 = (x-1)(x+6)$$

$$2x+6 = x^2 + 5x - 6$$

This is a quadratic equation, so set it equal to 0 to try and solve it by factoring.

$$0 = x^2 + 5x - 2x - 6 - 6$$

$$0 = x^2 + 3x - 12$$

Nuts, $x^2 + 3x - 12$ isn't factorable, so you'll either have to complete the square or apply the quadratic formula to get the solution. (I usually use the quadratic formula, so it doesn't feel like I memorized it for nothing.)

$$x = \frac{-b \pm \sqrt{b^2 - 4ac}}{2a} = \frac{-3 \pm \sqrt{3^2 - 4(1)(-12)}}{2(1)} = \frac{-3 \pm \sqrt{9 + 48}}{2} = \frac{-3 \pm \sqrt{57}}{2}$$

Okay, so that wasn't the most attractive problem in the world, but did you notice that the tough part was solving the quadratic equation, not dealing with the fractions? Here's even better news: you don't have to test these (horribly disgusting and unforgivably irrational) answers by plugging them into the original equation. Unlike multiplying by a least common denominator, cross multiplication won't introduce false solutions.

You've Got Problems

Problem 2: Solve the equation $\dfrac{3x-2}{x} = \dfrac{2x}{x+1}$.

Investigating Variation

There is an undeniable cause-and-effect link between many things in life. In some cases, an increase in one event causes an increase in another; for example, the more french fries you eat at a fast food restaurant, the more your waist will expand. Here's another one: the faster you drive, the more likely you are to get a speeding ticket.

On the other hand, there are also events that are paired up in the opposite (or inverse) way, an increase in one leads to a decrease in the other. I was once told that the number of keys you carry around at your job is inversely related to how successful you are at work. According to this theory, the boss only has one or two keys that open everything, but the worker bees have to tote around heavy key rings like zookeepers. How about this inverse relationship: the more successful you are at politics, the less honest you probably are as a person. (Ooh, a cheap shot!)

These relationships describe the variation between two values, and in this section, you'll learn to express them mathematically using rational equations.

Direct Variation

Direct variation is the relationship I described in which an increase in one value corresponds with an increase in the other, like in the statement "The longer you study for a test, the higher your score will be." In this case, an increase in study hours should correlate with an increase in test score.

Mathematically speaking, if x and y vary directly, then y is exactly k times as big as x. So, $y = k \cdot x$, where k is a number called the *constant of proportionality*. The k-value in the formula $y = k \cdot x$ describes how x and y are related, so your first job in a direct variation

Talk the Talk

If the variables x and y exhibit **direct variation**, then y is exactly k times as large as x; k is called the **constant of proportionality**. Direct variation is also called proportional variation.

problem is to calculate the constant of proportionality. You do that by dividing the two values that are directly related. So, if y varies directly with x, then $\frac{y}{x} = k$. Once you find k, you can use that equation to find any missing value.

Example 3: Assume that the measurements of the Statue of Liberty's facial features vary directly with the measurements of my facial features. The statue's nose is 54 inches long, whereas my nose is a (comparatively smaller) 2.375 inches, as illustrated in Figure 18.2. If Liberty's right eye measures 30 inches across, calculate the distance across my right eye. (Round all calculations to the thousandths place.)

Figure 18.2

The Statue of Liberty may have a large nose, but my nose is also pretty substantial, and I have no pointy hat to draw attention away from it.

Solution: Divide the length of the statue's nose by my nose length to determine the constant of proportionality. Use a calculator unless you're really excited about dividing by hand.

$$k = \frac{54}{2.375} \approx 22.737$$

Now that you know the value of k, you can calculate the width of my eye using the equation $\frac{y}{x} = k$. Plug the statue's eye measurement into the same variable you plugged her nose measurement into (y).

$$\frac{30}{x} = 22.737$$

Rewrite 22.737 as $\frac{22.737}{1}$ to create a proportion and cross multiply to solve for x.

$$\frac{30}{x} = \frac{22.737}{1}$$

$$30 \cdot 1 = x(22.737)$$

$$\frac{30}{22.737} = x$$

$$1.319 \approx x$$

Therefore, my eye is approximately 1.319 inches across. Of course, simply measuring my eye would have been easier for me than undertaking all that algebra, but then again, you don't have ready access to my eye, so algebra is your only option. I'm not about to start loaning my eye to people just so they can get the answer quicker. How could I be sure people would mail it back to me?

You've Got Problems

Problem 3: Assume x and y vary proportionally. If $x = 12$ when $y = 15$, what is the value of y when $x = 35$?

Indirect Variation

Indirect (or *inverse*) *variation* occurs when an increase in one quantity leads to a decrease in the other. Direct and inverse variation are very closely related. In direct variation, the quotient of the two values remains constant $\left(\frac{y}{x} = k\right)$, whereas inverse variation features a constant product ($xy = k$). Once again, your first job in an inverse variation problem will be to calculate k and then use it in the equation $xy = k$ to calculate missing values.

Talk the Talk

If two quantities, x and y, exhibit **indirect** (or **inverse**) **variation**, then their product remains constant even as the values of x and y change: $xy = k$. This means that as one quantity grows by a factor of n, the other shrinks by a factor of $\frac{1}{n}$.

Critical Point _____

In *The Simpsons* episode "HOMR," the reason for Homer's lack of intellectual acuity is revealed (in other words, we find out why he's such a dummy): there's a crayon stuck in his brain that's been lodged there since childhood. Doctors remove it, and suddenly he becomes a brainiac. However, it turns out to be a mixed blessing, because people shun the new, smarter Homer.

Grieved, he goes to talk to his superintelligent daughter Lisa, who affirms that his worst fears have been realized. "Dad, as intelligence goes up, happiness often goes down. In fact, I made a graph," she says, and holds up Figure 18.3. "I make a lot of graphs," she sighs sadly.

This is the graph of an inverse variation. Notice that as x gets bigger (and you get smarter), y gets smaller (the graph gets closer and closer to $y = 0$, the x-axis), meaning happiness decreases.

Inversely, the smaller your intelligence (the closer x is to the origin), the higher the graph, indicating great happiness.

Who says you couldn't learn something from watching television? In fact, *The Simpsons* television program often includes mathematical jokes. Check out Dr. Sarah J. Greenwald and Dr. Andrew Nestler's website that chronicles these (often subtle) references at www.simpsonsmath.com.

Figure 18.3

Lisa has bad news for her newly intelligent father. With great intelligence comes great unhappiness (and vice versa).

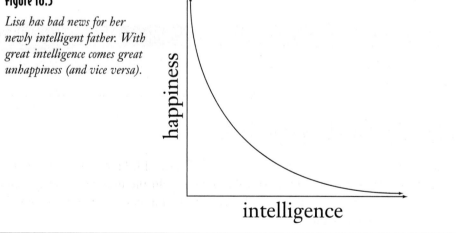

Example 4: Suppose that the number of times Oscar brushes his teeth in a year varies inversely with the total number of cavities he'll get that year. In 2003, he brushed his teeth a total of 48 times and got four cavities. If he plans to brush his teeth 140 times this year, how many cavities should he expect? (Round your answer to the nearest whole number.)

Solution: Let t equal the total number of times Oscar brushes his teeth in a year and c equal his total number of cavities for that year. The problem tells you that t and c vary inversely, so their product is a constant: $t \cdot c = k$. In the year 2003, $t = 48$ and $c = 4$; substitute those values into the equation to find k.

$$48 \cdot 4 = k$$

$$192 = k$$

Now that you know $k = 192$, it's time for new values of t and c. This year $t = 140$; calculate the corresponding value of c.

$$t \cdot c = k$$

$$140 \cdot c = 192$$

$$c = \frac{192}{140}$$

$$c \approx 1.371$$

Round your answer to the nearest whole number: $c = 1$. Oscar should expect only one cavity this year.

You've Got Problems

Problem 4: Assume x varies inversely with y. If $x = 9$ when $y = 75$, what is the value of x when $y = 2$?

Solving Rational Inequalities

Take a moment to review the last section of Chapter 13. Here's the process in a nutshell: factor the quadratic, find critical numbers, split the number line into intervals based on those critical numbers, and then test the intervals to determine which are solutions of the inequality.

You'll use a very similar process to solve rational inequalities. Actually, to be perfectly honest, the process is *exactly* the same. Chapter 13 defines a critical number as an x-value that either makes an expression equal zero or causes it to be undefined. Back then, however, nothing made the expressions undefined. Now that fractions have entered the picture, things will change a bit. Values that make the denominator of a fraction zero also make the function undefined.

If you follow these steps, solving rational inequalities is a piece of cake:

1. **Rearrange the inequality so that only zero remains on the right side.** This means you should add and subtract to move all of the terms on the right side to the left.

2. **Create one fraction on the left side of the inequality.** Use common denominators to combine multiple terms into a single fraction.

3. **Factor the numerator and denominator, if possible.** This makes finding critical numbers extremely easy.

4. **Set each factor in the numerator equal to zero and solve.** Mark these critical numbers on the number line using an open dot (if the inequality symbol is < or >) or a closed dot (if the inequality symbol is ≤ or ≥).

5. **Set each factor in the denominator equal to zero and solve.** The solutions are also critical numbers, and should always be marked on the number line with an open dot, because zeros in the denominator make the fraction undefined. (Remember, you're not allowed to divide by zero.)

6. **Choose test points to identify solution intervals.** The critical numbers split the number line into intervals. Choose one test value from each interval to plug into the inequality. True inequality statements indicate that the interval is part of the solution.

Be extra careful with the dots you place on the number line. If you use the wrong dot, you'll get the inequality signs in your answer wrong and the graph will be inaccurate.

Example 5: Solve the inequality $\dfrac{2x+5}{x+4} \geq -x$ and graph the solution.

Solution: Start by moving $-x$ to the left side of the inequality by adding x to both sides—the right side should only contain a zero.

$$\frac{2x+5}{x+4} + x \geq 0$$

Combine the terms on the left side of the inequality into a single fraction. The least common denominator of $\dfrac{2x+5}{x+4}$ and $\dfrac{x}{1}$ is $x + 4$, so multiply the numerator and denominator of $\dfrac{x}{1}$ by the LCD and combine the fractions.

$$\frac{2x+5}{x+4}+\left(\frac{x+4}{x+4}\right)\cdot\frac{x}{1}\geq 0$$

$$\frac{2x+5}{x+4}+\frac{x^2+4x}{x+4}\geq 0$$

$$\frac{2x+5+x^2+4x}{x+4}\geq 0$$

$$\frac{x^2+6x+5}{x+4}\geq 0$$

Factor the numerator.

$$\frac{(x+1)(x+5)}{x+4}\geq 0$$

Set each factor of the numerator equal to 0 and solve to get two critical numbers.

$$x+1=0 \qquad \text{or} \qquad x+5=0$$
$$x=-1 \qquad\qquad\qquad x=-5$$

Mark $x = -1$ and $x = -5$ on the number line using *solid* dots, as illustrated by Figure 18.4. Generate the final critical number by setting the denominator equal to 0.

$$x+4=0$$
$$x=-4$$

All critical numbers that come from the denominator are marked with open dots, as you can see in Figure 18.4.

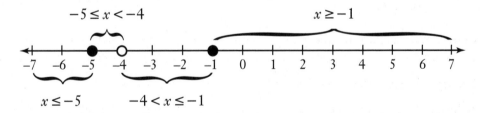

Figure 18.4

The critical numbers x = –5, –4, and –1 split the number line into four intervals.

Choose test points from each interval (I suggest $x = -6$, $x = -4.5$, $x = -2$, and $x = 0$) and plug them into the inequality. Use the factored, single-fraction version of the inequality to make things easier on yourself.

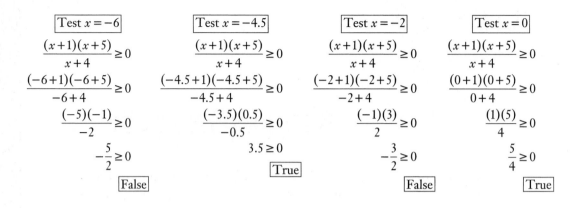

The test values $x = -4.5$ and $x = 0$ make the inequality true, so the solution of the inequality is made up of the intervals that contain them: $-5 \le x < -4$ or $x \ge -1$. Darken those intervals on the number line to create the graph, as illustrated by Figure 18.5.

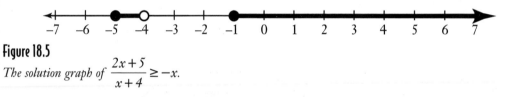

Figure 18.5

The solution graph of $\dfrac{2x+5}{x+4} \ge -x$.

You've Got Problems

Problem 5: Solve the inequality $\dfrac{x+7}{x-2} < 3$ and graph the solution.

The Least You Need to Know

♦ Multiplying all of the terms in a rational equation by the least common denominator eliminates the fractions.

♦ Cross multiplication eliminates the fractions in a proportion.

♦ If x varies *directly* with y, then $\dfrac{y}{x} = k$; if x varies *inversely* with y, then $xy = k$.

♦ Critical numbers are values that make an expression equal zero or cause it to be undefined.

Part 7

Wrapping Things Up

No algebra book is complete without a chapter on word problems. It would be like a dentist appointment without getting your teeth drilled, or an eye exam without a glaucoma-detecting burst of air shot into your eyeball. Word problems are a necessary evil of algebra, jammed in there to show you that you can use algebra in "real life." Once you persevere through the word problems, you'll find a comprehensive review of all the skills and concepts discussed in this book, so that you can practice to your heart's content.

Chapter 19

Whipping Word Problems

In This Chapter

- ◆ Calculating simple and compound interest
- ◆ Solving geometric measurement problems
- ◆ Determining distance and rate of travel
- ◆ Cracking problems with combinations and mixtures

The mythological Greek gods handed out very creative punishments. Take, for example, the extraordinarily unpleasant fate of Sisyphus. The guy was no peach, a cruel and heartless king too sly for his own good. When Death itself, in the form of Hades, came to claim him, he managed to shackle Death and hold it captive. Eventually, people started to notice that no one was dying ("Hey, Gary, sorry about running you over with my chariot this morning—I was changing radio stations. No offense, but you don't look so good decapitated") and Sisyphus's luck ran out.

As punishment for his gutsy kidnapping, he was doomed to roll a large boulder up a steep mountain in the land of the dead for all eternity. That's pretty nasty, but it's not the worst part. After hours and hours of hard labor, moving the mammoth rock slowly, gaining ground at an agonizingly slow rate, and straining every muscle in his body, just before reaching the zenith of the cliff and accomplishing his task, the boulder would roll all the way back down the hill. Time to start all over again.

I can't imagine such a depressing fate, to spend all of eternity forced to do something that is inherently painful and pointless, knowing that no matter how hard you try, you'll never be able to accomplish your task. However, thousands of students every day engage in their daily battle as Sisyphus. Each morning, they walk up to the giant cliff that is algebra and a massive boulder representing word problems.

Even though they are not doomed to work word problems unsuccessfully for all eternity, they lack one thing Sisyphus doesn't: direction. At least he knows *what he's supposed to do*. I have seen lots of students who stare open-mouthed at word problems with great, gleaming eyes, wet with unshed tears, who all say the same thing: "I don't even know where to start! How am I supposed to do these problems if I can't even figure out the first step?" (Of course, there's more cursing when *they* say it, but you get the idea.)

In this chapter, I'm going to risk incurring the wrath of the mythological algebraic deities and free you from your doom. I'll introduce you to four of the most common types of word problems and provide a plan of attack for each one. That way, instead of living in fear, you can stare those problems down and calmly retort (in the voice and comedic timing of a big-screen action hero), "Let's rock." You'll feel much boulder. (Horrible puns definitely intended.)

Interest Problems

There are three good reasons to deposit your life savings in a bank account, rather than hide it in your closet or mattress:

1. A bank is safer, and if your money is stolen, there are usually federal laws that insure your investment.

2. A bank affords you the unique opportunity to write with pens chained to desks. Even though your money allows banks to rake in the dough, for some reason they are very adamant that you not accidentally take their pens.

3. You earn interest on your money without having to exert any effort at all.

Interest is a great thing. It's free money that you earn just by keeping your money in a safe place. You may be asked to solve problems in which you calculate the interest earned by some initial investment (which is called the *principal*) after a period of time has passed. There are two major types of interest problems you may be asked to solve: simple interest and compound interest.

Simple Interest

If your money grows with simple interest, you're basically earning a small percentage of your initial investment each year as interest. For instance, if the principal of an account is $100 and your annual interest rate is 6.75%, at the conclusion of every year, you will have earned an additional $6.75 (since $6.75 is 6.75% of $100).

Here's the bad news: even though your account will grow slightly every year, you only earn interest on the initial investment, no matter how long you've had an active bank account or how much interest that money has accrued.

The formula for calculating simple interest is $i = prt$, where p is the principal, r is the annual interest rate (expressed as a decimal), and i is the interest you have earned after the money has been invested for t years.

Example 1: You were a very thrifty and money-savvy child. Instead of spending the money from the tooth fairy, you invested that cash as one lump sum of $32.00 in a bank account with a fixed annual interest rate of 7.75%. What is the balance of the account exactly 30 years later?

Solution: To calculate the balance, you'll add the interest earned to the principal. Of course, you still need to figure what that interest is. Use the formula $i = prt$, where $p = 32$, $r = 0.0775$ (the decimal equivalent of 7.75%), and $t = 30$.

$$i = prt$$
$$= (32)(0.0775)(30)$$
$$= \$74.40$$

You earned $74.40 in interest over that 30-year period, so add that to the initial investment to calculate the account balance.

$$\text{balance} = \text{principal} + \text{interest earned}$$
$$= \$32 + \$74.40$$
$$= \$106.40$$

Compound Interest

Most banks don't use simple interest; the more money you deposit, the more money they can potentially make, so they want to encourage you to deposit as much as possible into your account. One way they do this is via *compound interest*, in which you earn money based on your original principal *and* the interest you accrue.

Let's say you deposit $100 in an account where the interest is compounded annually at a rate of 6.0%. At the end of the first year, you will have a balance of $106, just like you would with simple interest. However, at the end of the second year, you'll earn 6.0% interest on the new balance of $106, not just the original balance of $100.

Talk the Talk

If your bank account accrues **compound interest**, then you earn interest based upon your entire balance, rather than just the initial investment.

Critical Point

The more times the interest in your account is compounded, the more money you'll earn. The best possible scenario would be continuously compounding interest, which compounds an infinite number of times each year. That's a little tricky, so you'll have to wait for precalculus to learn how that works.

The interest for most bank accounts is compounded more often than once a year. Whether it is compounded weekly (52 times a year), monthly (12 times a year), or quarterly (4 times a year) can make a considerable difference to your balance.

The compound interest formula is slightly more complicated than the formula for simple interest:

$$b = p\left(1 + \frac{r}{n}\right)^{nt}$$

In this formula, p is once again the principal investment, r is the annual interest rate in decimal form, and t is the length of time the money is invested. There are two new variables: n, the number of times interest is compounded in one year, and b, the balance of the account.

Example 2: How much more money would you make over an 18-month period if you invested $3,000 in a savings account with a 6.25% annual interest rate that was compounded monthly rather than quarterly? (To keep answers consistent, round all decimals to seven decimal places as you calculate.)

Solution: You'll have to calculate two separate balances, one with $n = 12$ for monthly compounding, and one with $n = 4$ for quarterly interest compounding. The other variables will match for both problems: $p = 3,000$, $r = 0.0625$, and $t = 1.5$. Be careful, the variable t is measured in years, not months. Because 18 months is exactly a year and a half, $t = 1.5$.

Calculate the balance when $n = 12$.

$$b = 3,000\left(1 + \frac{0.0625}{12}\right)^{(12)(1.5)}$$
$$= 3,000(1 + 0.0052083)^{18}$$
$$= 3,000(1.0980173)$$
$$= 3,294.0519$$

Kelley's Cautions

Notice that the compound interest formula gives you the total *balance*, whereas the simple interest formula gives you the *interest*—you had to add the principal to the interest in Example 1 in order to calculate the balance.

Round your answer to the nearest penny (to two decimal places): $3,294.05. Now calculate the balance when $n = 4$.

$$b = 3,000\left(1 + \frac{0.0625}{4}\right)^{(4)(1.5)}$$
$$= 3,000(1 + 0.015625)^{6}$$
$$= 3,000(1.0974893)$$
$$= 3,292.47$$

Subtract the two balances to find the total difference: $3,294.05 − $3,292.47 = $1.58. Sure, $1.58 isn't a huge difference, but the larger the principal and the longer you invest, the larger that difference will grow.

Kelley's Cautions

In both simple and compound interest problems, t must be measured in years. Therefore, you should set $t = 2$ (not $t = 24$) for a 24-month investment.

You've Got Problems

Problem 1: Calculate the balance of an account if a $5,000 principal investment earns:

   (a) Simple interest at an annual rate of 8.25% for 20 years.

   (b) Interest compounded weekly at an annual rate of 8.25% for 20 years.

Round all calculations to seven decimal places.

Area and Volume Problems

Algebra teachers love fractals. I don't think I've ever met a math teacher who didn't have a fractal poster proudly displayed in his classroom, office, or wallet. (In case you don't know, fractals are geometric shapes that look the same at any magnification—they have the same shape no matter how far you zoom in or out.) Algebra teachers love geometric shapes so much, in fact, that they often ask you questions about them before you've taken a single geometry class. Don't be surprised if you see word problems that refer to the following geometric concepts:

- **Area:** The amount of space covered by a two-dimensional object; the amount of carpet you'd need to cover a floor represents the floor's area.

- **Perimeter:** The distance around a two-dimensional object; the amount of fence you'd need to surround your yard equals the perimeter of your yard. (The perimeter of a circular object is called the *circumference*.)

- **Volume:** The amount of three-dimensional space inside an object; the amount of liquid inside a soda can represents the can's volume.

- **Surface area:** Measures the amount of "skin" needed to cover a three-dimensional object, neglecting its thickness; the amount of leather used to cover a basketball represents the surface area of the ball.

There are lots of geometric formulas, most of which you'll learn in a geometry class, but you should become familiar with the formulas in Table 19.1.

Table 19.1 Basic Geometric Formulas

Description	Formula	Variables
Area of a rectangle	$A = l \cdot w$	l = length, w = width
Perimeter of a rectangle	$P = 2l + 2w$	l = length, w = width
Area of a circle	$A = \pi r^2$	r = radius
Circumference of a circle	$C = 2\pi r$	r = radius
Volume of a rectangular solid	$V = l \cdot w \cdot h$	l = length, w = width, h = height
Volume of a cylinder	$V = \pi r^2 h$	r = radius, h = height
Surface area of a cube	$S = 6l^2$	l = length of side

If you know the correct formulas, geometric word problems are pretty simple.

Example 3: (Inspired by *National Lampoon's Christmas Vacation*) Uncle Eddie is nervous about sledding with Clark W. Griswold, thanks to the rectangular metal plate in his head. (As it is, every time his wife uses the microwave, he forgets who he is for a half hour.) During the operation to implant the plate, right before the anesthetic kicked in, Eddie remembers hearing the doctor say that the plate's width is 3 centimeters shorter than its length. If the area of the plate is 54 cm², find its dimensions.

Solution: You know nothing at all about the length of the rectangle, but you do know that the width equals the length minus 3 centimeters, so write that expression algebraically: $w = l - 3$. The formula for the area of a rectangle is $A = l \cdot w$. Substitute $A = 54$ and $w = l - 3$ into the formula.

$$A = l \cdot w$$
$$54 = l(l - 3)$$
$$54 = l^2 - 3l$$

Set the quadratic equation equal to zero and solve by factoring.

$$0 = l^2 - 3l - 54$$
$$0 = (l - 9)(l + 6)$$
$$l - 9 = 0 \qquad l + 6 = 0$$
$$l = 9 \quad \text{or} \quad l = -6$$

One of these answers doesn't make sense. The length of an object must *always* be positive, so throw out $l = -6$. Therefore, the length of the rectangle must be 9 cm. Remember that $w = l - 3$; substitute $l = 9$ into that equation to determine the width of the rectangle.

$$w = l - 3 = 9 - 3 = 6 \text{ cm}$$

You've Got Problems

Problem 2: The height of a certain cylinder is exactly twice as large as its radius. If the volume of the cylinder is 36π in³, what is the radius of the cylinder?

Speed and Distance Problems

Have you ever heard of a word problem like this one? "Train A heads north at an average speed of 95 miles per hour, leaving its station at the precise moment another train, Train B, departs a different station, heading south at an average speed of 110 miles per hour. If these trains are inadvertently placed on the same track and start exactly 1,300 miles apart, how long until they collide?"

If that problem sounds familiar, it's probably because you watch a lot of television (like me). Whenever TV shows mention math, it's usually in the context of a main character trying (but failing miserably) to solve the classic "impossible train problem." I have no idea why that is, but time and time again, this problem is singled out as the reason people hate math so much.

Surprisingly, this problem is not so hard. Like most distance and rate of travel problems, it only requires one simple formula: $D = r \cdot t$. Distance traveled is equal to the rate of speed (r) multiplied by the time (t) you traveled that speed. What makes most distance and rate problems tricky is that you usually have two things traveling at once, so you need to use the formula twice at the same time. In this problem, you'll use it once for Train A and once for Train B.

To keep things straight in your mind, you should use little descriptive subscripts. For example, use the formula $D_A = r_A \cdot t_A$ for Train A's distance, speed, and time values, and use the formula $D_B = r_B \cdot t_B$ for Train B.

Example 4: Train A heads north at an average speed of 95 miles per hour, leaving its station at the precise moment another train, Train B, departs a different station, heading south at an average speed of 110 miles per hour. If these trains are inadvertently placed on the same track and start exactly 1,300 miles apart, how long until they collide?

Solution: Two trains means two distance formulas: $D_A = r_A \cdot t_A$ and $D_B = r_B \cdot t_B$. Your first goal is to plug in any values you can determine from the problem. Train A travels 95 mph, so $r_A = 95$; similarly, $r_B = 110$.

The problem indicates that the trains leave at the same time, which means their travel times match exactly. Therefore, instead of denoting their travel times as t_A and t_B (which suggests they are different), write them both as t (which suggests they are equal). Right now, the two distance formulas look like this:

$$D_A = r_A \cdot t_A \qquad D_B = r_B \cdot t_B$$
$$D_A = 95t \qquad D_B = 110t$$

Here's the tricky step. The trains are heading toward one another on a track that's 1,300 miles long. Therefore, they will collide when, together, both trains have traveled a total of 1,300 miles. Because Train B is traveling faster, it will also travel further than Train A, but that doesn't matter. You don't even have to figure out how far each train will go. All that matters is that when $D_A + D_B = 1,300$, it's curtains. Luckily, you happen to know what D_A and D_B are ($95t$ and $110t$, respectively) so plug those into the equation and solve.

$$D_A + D_B = 1{,}300$$
$$95t + 110t = 1{,}300$$
$$205t = 1{,}300$$
$$t = \frac{1{,}300}{205}$$
$$t \approx 6.341 \text{ hours}$$

The trains will collide in approximately 6.341 hours. The dining cars will be unavailable at that time.

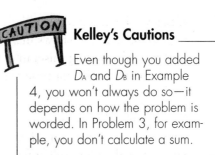

Kelley's Cautions

Even though you added D_A and D_B in Example 4, you won't always do so—it depends on how the problem is worded. In Problem 3, for example, you don't calculate a sum.

You've Got Problems

Problem 3: Dave rode his bike from home to a convenience store at an average speed of 17 mph, and the trip took 1.25 hours. However, as he pulled up to the store, he rode over some glass, causing both tires to go flat. Because of this rotten luck, he had to push his bike back home at an average speed of 3 mph. How long, in hours, did the trip home take? Round your answer to two decimal places.

Mixture and Combination Problems

The final type of word problem involves mixing two or more things together. What's actually being mixed doesn't matter, because the process is the same no matter what the ingredients are.

Your job will be to multiply the amount of each individual ingredient by some quantity that describes it, like its concentration or weight, and then add those products, like so:

$$\boxed{\text{ingredient 1}} \cdot \boxed{\text{its value}} + \boxed{\text{ingredient 2}} \cdot \boxed{\text{its value}} = \boxed{\text{total amount}} \cdot \boxed{\text{its value}}$$

Like the individual ingredients, the total amount is multiplied by its descriptive value.

Example 5: A five-pound bag of trail mix consists of 20% raisins by weight. The contents of the bag are mixed together with a three-pound bag of trail mix that consists of 10% raisins by weight. What percentage of the resulting mixture is raisins?

Kelley's Cautions

If the descriptive value for an ingredient is a percentage, convert it to a decimal for the mixture equation.

Solution: Let's be honest. If you weren't in an algebra class, the correct answer to this question would be "Who cares? I don't even like raisins," but that won't get you any partial credit, and it usually makes the teacher kind of angry.

Let the five-pound bag be ingredient 1; its descriptive value is the percentage of raisins expressed as a decimal: 20% = 0.20. Ingredient 2 is the three-pound bag, and its descriptive value is 10% = 0.10. The total combined weight of both bags of trail mix is 3 + 5 = 8 pounds, but you don't know the weight of the raisins in the final mixture, so let that equal x. Create the mixture equation using these values and solve for x.

$$5(0.20) + 3(0.10) = 8(x)$$
$$1 + 0.3 = 8x$$
$$1.3 = 8x$$
$$\frac{1.3}{8} = x$$
$$0.1625 = x$$

Convert the decimal into a percent by moving the decimal point two places to the right: x = 16.25%. The giant, unappetizing 8-pound barrel of trail mix is 16.25% raisins by weight. Hope you're hungry.

You've Got Problems

Problem 4: A 10-gallon fish tank currently contains 7 gallons of water with a 1.2% saline concentration. What must the saline concentration of the final three gallons of water be to fill the tank and reach an overall saline concentration of 2%? Report the percentage accurate to two decimal places.

The Least You Need to Know

◆ Compound interest allows you to earn money based on the principal and interest earned to date, whereas simple interest is only earned on the principal.

◆ Geometric word problems require you to use algebra to calculate area, perimeter, volume, or surface area.

◆ Distance traveled is equal to the rate of travel multiplied by the time traveled.

◆ In mixture problems, you multiply the amount of each ingredient by some descriptive value, like its concentration.

20

Final Exam

In This Chapter

- Measuring your understanding of all major algebra topics
- Practicing your skills
- Determining where you need more practice

Nothing helps you understand algebra like good, old-fashioned practice, and that's the purpose of this chapter. You can use it however you like, but I suggest one of the following three strategies:

1. As you finish reading each chapter, skip back here and work on the practice problems from that chapter.

2. If you're using this book as a refresher for an algebra class you've already taken, complete this chapter before you start reading the book. Then go back and work through the chapters containing the problems you struggled with.

3. Save this chapter until the end, and use it to see how many of the topics you've mastered once you're done reading this book.

Because these problems are only meant for practice, not to actually teach the concepts, only the answers are given at the end of the chapter, without explanation or justification (unlike the problems in the "You've Got Problems" sidebars throughout the book). However, these practice problems are designed to mirror those examples, so look at the in-chapter practice problems if you forgot something.

Are you ready? There's a lot of practice ahead of you. Some of the problems below have multiple parts, so there are over 110 practice problems here! Don't worry—no one said you have to do them all at once.

Chapter 1

1. Identify all the categories the number $0.6\overline{21}$ (or 0.621212121...) belongs to.

2. Simplify $2 + (-3) - (-5) - (+7) + (+1)$.

3. Simplify.

    (a) $6 \times (-5)$

    (b) $-14 \div (-2)$

4. Simplify.

    (a) $4 + [3 - (8 + 2)]$

    (b) $|6 - 2(5 + 4)|$

5. Name the mathematical property that guarantees each of the below statements is true.

    (a) $3 + (1 + 4) = (3 + 1) + 4$

    (b) $\frac{1}{2} \cdot 2 = 1$

    (c) $-8 + 0 = -8$

    (d) $2 \times 3 \times 5 = 3 \times 5 \times 2$

Chapter 2

6. Simplify the fraction $\frac{40}{64}$.

7. Rewrite the fractions so that they contain the least common denominator:

    $\frac{1}{2}, \frac{9}{4}, \frac{5}{6}$.

8. Simplify.

   (a) $\dfrac{2}{5} + \dfrac{7}{3} - \dfrac{1}{2}$

   (b) $\dfrac{5}{7} \cdot \dfrac{1}{10} \cdot \dfrac{2}{3}$

   (c) $\dfrac{6}{5} \div \dfrac{1}{3}$

Chapter 3

9. Translate the following statements into mathematical expressions.

   (a) Seventeen less than a number

   (b) The product of a number and one more than three times that number

10. Evaluate the expressions.

    (a) $(-2)^3$

    (b) 5^4

11. Simplify the expression $\left(\dfrac{x^3 y^{-2}}{x^{-1}} \right)^4$.

12. Write the numbers in scientific notation.

    (a) 0.00000679

    (b) 23,400,000,000,000,000,000,000,000

13. Apply the distributive property: $-3x(4x^5 + 7x - 9)$.

14. Simplify the expression: $10 - (12 - 4 \times 2)^2$.

15. Evaluate the expression $3x(xy - y^2)$ if $x = 2$ and $y = -3$.

Chapter 4

16. Solve the equations.

    (a) $3 + w = 12$

    (b) $-\dfrac{7}{5}x = \dfrac{2}{15}$

    (c) $7y - 19 = 2y - 4$

(d) $4(x - 5) + 8 = -28$

(e) $2|x - 3| + 5 = 13$

17. Solve the equation $5x - 4y = 20$ for y.

Chapter 5

18. Identify the points indicated on the coordinate plane below.

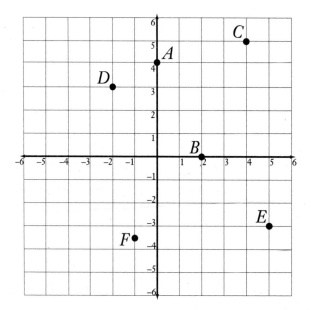

19. What are the coordinates of the intercepts of the linear equation $4x - 3y = 8$?

20. Which of the following equations has the following graph?

(A) $2x + 3y = 6$

(B) $-2x + 3y = 6$

(C) $2x - 3y = 6$

(D) $2x + 3y = -6$

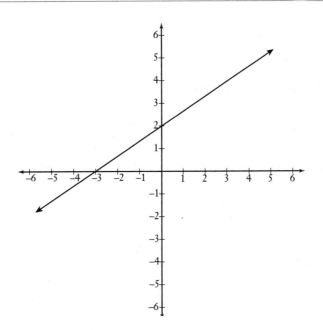

21. Calculate the slope of the line that passes through the points (4,3) and (–9,2).

22. What coordinate pair represents the vertex of the graph of the equation $y = -|x-4|-5$?

Chapter 6

23. Determine the equation of the line with slope $m = -\dfrac{2}{3}$ that passes through the point (–1,2) and write it in standard form.

24. Determine the slope and the y-intercept of the line with equation $5x - 3y = -9$.

25. Write the equation of the line that passes through the points (–3,7) and (4,1) in standard form.

26. Write the equation of line m in standard form if m passes through the point (7,0) and is perpendicular to the line $2x + y = -6$.

Chapter 7

27. Solve the inequality $14 - 3(y + 5) > 8$.

28. Which of the following is the graph of the inequality $-3 < 2x + 5 \leq 1$?

(A)

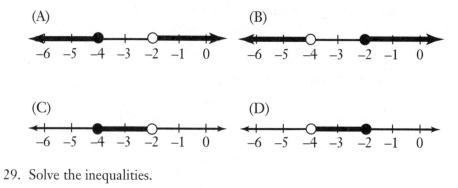

(B)

(C)

(D)

29. Solve the inequalities.

    (a) $|2x - 1| > 0$

    (b) $|x + 4| \leq 2$

30. Write the equation of the inequality graphed below in standard form.

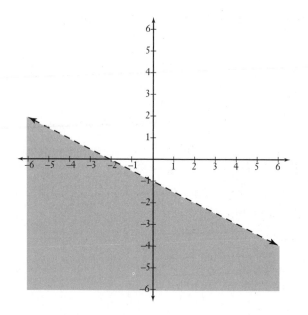

Chapter 8

31. Solve the system by graphing the equations.

$$\begin{cases} 2x - y = 7 \\ x + 3y = -7 \end{cases}$$

32. Solve the system using substitution.

$$\begin{cases} 3x + y = 1 \\ 4x + 3y = -2 \end{cases}$$

33. Solve the system using elimination.

$$\begin{cases} 4x + 3y = -1 \\ -2x + 15y = 6 \end{cases}$$

34. Describe the solution of the system of equations.

$$\begin{cases} x - 3y = -4 \\ 3x - 9y = -11 \end{cases}$$

35. Graph the solution of the system of inequalities.

$$\begin{cases} y \le 4x - 3 \\ y > -\dfrac{1}{3}x + 1 \end{cases}$$

Chapter 9

36. If $A = \begin{bmatrix} 4 & -1 \\ 3 & 0 \\ -2 & 9 \end{bmatrix}$ and $B = \begin{bmatrix} -3 & 1 \\ 6 & -2 \\ 12 & 5 \end{bmatrix}$, calculate $2A - B$.

37. Calculate $\begin{bmatrix} 7 & -3 \\ -2 & 1 \end{bmatrix} \cdot \begin{bmatrix} -1 & 0 & 4 & -6 \\ -8 & 5 & 1 & 2 \end{bmatrix}$.

38. Evaluate the determinants.

 (a) $\begin{bmatrix} -6 & -5 \\ -1 & 9 \end{bmatrix}$

 (b) $\begin{bmatrix} 2 & -3 & -1 \\ 1 & -4 & 0 \\ 5 & 1 & 6 \end{bmatrix}$

39. Use Cramer's Rule to solve the system of equations.

$$\begin{cases} 9x + 2y = 7 \\ 36x - 3y = -16 \end{cases}$$

Chapter 10

40. Classify the polynomials.

   (a) $-7x^4 - 9$

   (b) $19x^3$

41. Simplify the expression: $3x^2(2xy + 8x - 3y) - 2xy(y^3 + 2x^2 - 7x)$.

42. Calculate the products and simplify.

   (a) $(2x^2y - 3xy^3)(xy + 7y^3)$

   (b) $(x - 3y)(x^2 + 2xy - y^2)$

43. Calculate the quotients.

   (a) $(x^4 + 3x^3 - x + 5) \div (x^2 - 2x + 1)$

   (b) $(x^4 - 9x^3 + 25x^2 - 26x + 17) \div (x - 5)$

Chapter 11

44. Factor the polynomials.

   (a) $12x^3y - 6x^2y^2 + 9x^5y^3$

   (b) $12x^2 + 18x - 10xy - 15y$

   (c) $16x^2 - 1$

   (d) $x^3 - 8$

   (e) $x^2 + 9x - 36$

   (f) $6x^2 - 17x - 3$

Chapter 12

45. Simplify the radicals.

   (a) $\sqrt[3]{-27x^{10}y^8}$

   (b) $\sqrt{75x^2y^3}$

46. Rewrite the expression $8^{7/3}$ as an integer.

47. Simplify the radical expressions, rationalizing if necessary.

    (a) $\sqrt{12x} - \sqrt{48x}$

    (b) $\left(\sqrt[3]{9x^2y}\right)\left(\sqrt[3]{6x^2y^2}\right)$

    (c) $\sqrt{35x^2y^7} \div \sqrt{7xy^8}$

48. Solve the equation: $\sqrt[3]{x+7} = 2$.

49. Simplify the expressions.

    (a) i^{11}

    (b) $\sqrt{-18x}$

50. Simplify the expressions.

    (a) $(3 - 2i) - 5(-2 + 3i)$

    (b) $(-3 + i)(5 - 6i)$

    (c) $(2 + i) \div (4 - i)$

Chapter 13

51. Solve the equation by factoring: $2x^2 - x = 15$.

52. Solve by completing the square: $3x^2 + 12x - 21 = 0$.

53. Solve using the quadratic formula: $2x^2 + 5x + 6 = 0$.

54. Without calculating them, determine how many real solutions the equation
    $9x^2 - 3x = -\dfrac{1}{4}$ has.

55. Solve the inequality and graph the solution: $x^2 + 2x - 24 \le 0$.

Chapter 14

56. Demonstrate that $x - 1$ is a factor of the polynomial $2x^3 - 11x^2 + 4x + 5$ and use that information to completely factor the trinomial.

57. If 8 is a root of the equation $x^3 - 5x^2 - 42x + 144 = 0$, use that information to find the other two roots.

58. Identify all four rational roots of the polynomial equation
    $4x^4 - 13x^3 - 37x^2 + 106x - 24 = 0$.

59. Find all the roots of the equation: $x^3 + 7x^2 + 19x + 45 = 0$.

Chapter 15

60. Calculate the following if $f(x) = x^3 + 4x - 1$ and $g(x) = x^2 + 6$.

   (a) $(f - g)(x)$

   (b) $(fg)(-2)$

61. Calculate the following if $h(x) = (x + 1)^2$ and $j(x) = x - 3$.

   (a) $h(j(x))$

   (b) $(j \circ h)(-1)$

62. If $f(x) = 6(x - 1) + 2$, find $f^{-1}(x)$.

63. Given $g(x)$ as defined below, evaluate $g(-2)$ and $g(3)$.

$$g(x) = \begin{cases} |2x + 3|, & x \leq -2 \\ 2x + 3, & -2 < x < 2 \\ -|2x + 3|, & x \geq 2 \end{cases}$$

Chapter 16

64. Sketch the function $f(x) = \sqrt{-x} + 2$ using transformations, and then verify the graph's accuracy by graphing the function again, this time by plotting points.

65. Examine the graph of $p(x)$ below. Is $p(x)$ a function? If so, is $p(x)$ one-to-one? Why or why not?

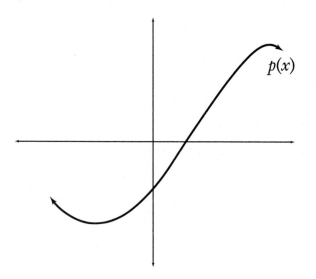

$p(x)$

66. Determine the domain and range of $k(x)$, graphed below.

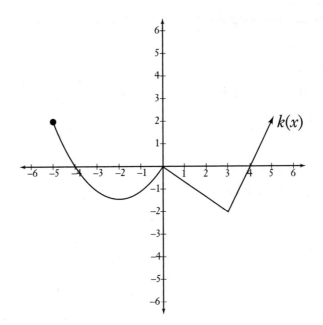

Chapter 17

67. Simplify the expression $\dfrac{3x^2 + 13x - 10}{2x^2 + 11x + 5}$.

68. Rewrite each as a single expression and simplify.

(a) $\dfrac{3x}{x^2 - 9x + 20} - \dfrac{2}{x - 5}$

(b) $\dfrac{x^2 y^3}{x^2 + 13x + 36} \cdot \dfrac{x^2 - 81}{x^4 y^2}$

(c) $\dfrac{x^2 - 6x - 27}{x^2 - 11x + 18} \div \dfrac{x^3 + 2x^2 - 3x}{x^2 - 2x}$

69. Simplify the below complex fraction.

$$\dfrac{\dfrac{6x + 4}{x^2}}{\dfrac{12x + 8}{x^3 - x}}$$

Chapter 18

70. Solve the equations.

   (a) $\dfrac{x}{2x^2 - 5x + 3} - \dfrac{x+5}{2x^2 + x - 6} = \dfrac{6}{x^2 + x - 2}$

   (b) $\dfrac{x+6}{18} = \dfrac{-3}{x-9}$

71. The variables x and y vary proportionally, and $x = 3$ when $y = 5$. Determine the value of x when $y = 16$.

72. Assume x varies inversely with y, and that $y = 35$ when $x = 5$. Find the value of y when $x = 25$.

73. Solve and graph the inequality: $\dfrac{3x+1}{x-4} \le 2.$

Chapter 19

74. How much simple interest will a principal investment of $9,500 earn in a savings account with a fixed annual interest rate of 4.95% if it is left alone for 35 years?

75. Exactly 20 years ago, a deposit was made into a savings account with a 6.8% annual interest rate compounded quarterly. If the current balance is $55,852.71, what was the principal investment?

76. You want to mark the perimeter of your rectangular yard with fencing, and you bought exactly the correct amount of fencing to accomplish the task: 362 feet. If the length of your yard is exactly 1 foot longer than twice its width, find the yard's length.

77. The hare wants a rematch. Learning from his past mistakes pitted against the tortoise, he will hold off on the victory laps until the race is over. If the tortoise gets a head start of 90 minutes and travels at an average velocity of 3 feet/minute, how long will it take the hare (who travels at an average velocity of 600 feet/minute) to overtake the tortoise?

78. Sixteen ounces of an apple-flavored drink containing 15% real apple juice are mixed with 64 ounces of another juice drink. If the resulting mixture is 12.6% apple juice, what concentration of juice did the 64-ounce drink contain?

Solutions

Chapter 1: (1) Positive, real, rational, complex (see Chapter 12); (2) –2; (3a) –30; (3b) 7; (4a) –3; (4b) 12; (5a) The associative property of addition; (5b) Multiplicative inverse property; (5c) Identity property for addition; (5d) Commutative property of multiplication.

Chapter 2: (6) $\dfrac{5}{8}$; (7) $\dfrac{6}{12}$, $\dfrac{27}{12}$, $\dfrac{10}{12}$; (8a) $\dfrac{67}{30}$; (8b) $\dfrac{1}{21}$; (8c) $\dfrac{18}{5}$.

Chapter 3: (9a) $x - 17$; (9b) $x(3x + 1)$; (10a) –8; (10b) 625; (11) $\dfrac{x^{16}}{y^8}$; (12a) 6.79×10^{-6}; (12b) 2.34×10^{25}; (13) $-12x^6 - 21x^2 + 27x$; (14) –6; (15) –90.

Chapter 4: (16a) $w = 9$; (16b) $x = -\dfrac{2}{21}$; (16c) $y = 3$; (16d) $x = -4$; (16e) $x = -1$ or 7; (17) $y = \dfrac{5}{4}x - 5$.

Chapter 5: (18) $A = (0,4)$, $B = (2,0)$, $C = (4,5)$, $D = (-2,3)$, $E = (5,-3)$, and $F = \left(-1, -3\dfrac{1}{2}\right)$ or $\left(-1, -\dfrac{7}{2}\right)$; (19) $(2,0)$ and $\left(0, -\dfrac{8}{3}\right)$; (20) B; (21) $\dfrac{1}{13}$; (22) $(4,-5)$.

Chapter 6: (23) $2x + 3y = 4$; (24) slope $= \dfrac{5}{3}$, y-intercept $= 3$; (25) $6x + 7y = 31$; (26) $x - 2y = 7$.

Chapter 7: (27) $y < -3$; (28) D; (29a) $x > \dfrac{1}{2}$ or $x < \dfrac{1}{2}$ (all real numbers except $x = \dfrac{1}{2}$); (29b) $-6 \le x \le -2$; (30) $x + 2y < -2$.

Chapter 8: (31) $(2,-3)$, according to the graph below;

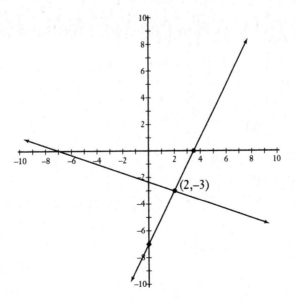

$(2,-3)$

(32) (1,–2); (33) $\left(-\dfrac{1}{2}, \dfrac{1}{3}\right)$; (34) Inconsistent (no solution); (35) See graph that follows.

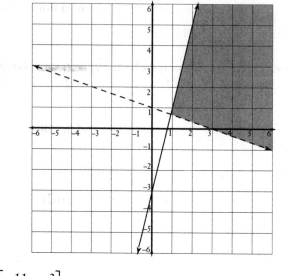

Chapter 9: (36) $\begin{bmatrix} 11 & -3 \\ 0 & 2 \\ -16 & 13 \end{bmatrix}$; (37) $\begin{bmatrix} 17 & -15 & 25 & -48 \\ -6 & 5 & -7 & 14 \end{bmatrix}$; (38a) –59; (38b) –51;

(39) $\left(-\dfrac{1}{9}, 4\right)$.

Chapter 10: (40a) Quartic binomial; (40b) Cubic monomial;

(41) $2x^3y + 24x^3 + 5x^2y - 2xy^4$; (42a) $2x^3y^2 + 11x^2y^4 - 21xy^6$;

(42b) $x^3 - x^2y - 7xy^2 + 3y^3$; (43a) $x^2 + 5x + 9 + \dfrac{12x - 4}{x^2 - 2x + 1}$;

(43b) $x^3 - 4x^2 + 5x - 1 + \dfrac{12}{x - 5}$.

Chapter 11: (44a) $3x^2y(4x - 2y + 3x^3y^2)$; (44b) $(6x - 5y)(2x + 3)$; (44c) $(4x + 1)(4x - 1)$; (44d) $(x - 2)(x^2 + 2x + 4)$; (44e) $(x + 12)(x - 3)$; (44f) $(6x + 1)(x - 3)$.

Chapter 12: (45a) $-3x^3y^2\left(\sqrt[3]{xy^2}\right)$; (45b) $5|xy|\sqrt{3y}$; (46) 128; (47a) $-2\sqrt{3x}$;

(47b) $3xy\sqrt[3]{2x}$; (47c) $\sqrt{\dfrac{5x}{y}}$; (48) $x = 1$; (49a) $-i$; (49b) $3i\sqrt{2x}$; (50a) $13 - 17i$;

(50b) $-9 + 23i$; (50c) $\dfrac{7}{17} + \dfrac{6}{17}i$.

Chapter 13: (51) $x=-\dfrac{5}{2}$, $x=3$; (52) $x=-2+\sqrt{11}$, $x=-2-\sqrt{11}$;

(53) $x=-\dfrac{5}{4}+\dfrac{\sqrt{23}}{4}i$, $x=-\dfrac{5}{4}-\dfrac{\sqrt{23}}{4}i$; (54) One (a double root);

(55) $-6\le x\le 4$, see graph below.

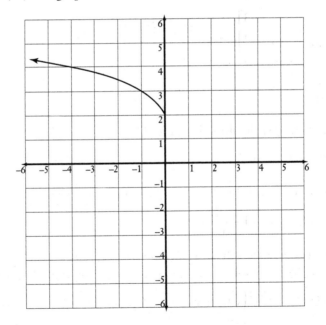

Chapter 14: (56) $(x-1)(2x+1)(x-5)$; (57) –6 and 3; (58) –3, $\dfrac{1}{4}$, 2, 4; (59) –5, $-1+2i\sqrt{2}$, $-1-2i\sqrt{2}$.

Chapter 15: (60a) $(f-g)(x)=x^3-x^2+4x-7$; (60b) $(fg)(-2)=-170$;

(61a) $h(j(x))=x^2-4x+4$; (61b) $(j\circ h)(-1)=-3$; (62) $f^{-1}(x)=\dfrac{1}{6}x+\dfrac{2}{3}$;

(63) $g(-2)=1$; $g(3)=-9$.

Chapter 16: (64) See graph below.

(65) $p(x)$ is a function because it passes the vertical line test, but is not one-to-one because it fails the horizontal line test; (66) Domain: $x\ge -5$, range: $y\ge -2$.

Chapter 17: (67) $\dfrac{3x-2}{2x+1}$; (68a) $\dfrac{x+8}{(x-5)(x-4)}$; (68b) $\dfrac{y(x-9)}{x^2(x+4)}$; (68c) $\dfrac{1}{x-1}$; (69) $\dfrac{x^2-1}{2x}$.

Chapter 18: (70a) $x = \dfrac{23}{14}$; (70b) $x = 0,3$; (71) $x = \dfrac{48}{5}$; (72) $y = 7$; (73) $-9 \le x < 4$, see graph below.

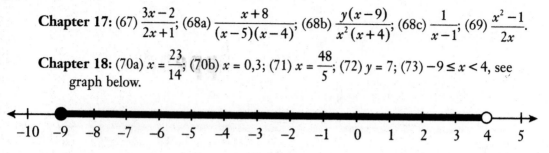

Chapter 19: (74) \$16,458.75; (75) \$14,500; (76) 121 feet; (77) The hare passes the tortoise after just $\dfrac{90}{199} \approx 0.452$ minutes; (78) 12%.

Solutions to "You've Got Problems"

Here are detailed solutions for all the "You've Got Problems" sidebars throughout the book. I suggest you turn back here to look only after you've tried your best and either arrived at an answer or are hopelessly stuck; you should find just enough information to get you through any important or tricky steps.

Don't just look at the problem and then flip back here to read the answer! Unless you actually *do the problem yourself first*, you'll never master the concept on your own.

Chapter 1

1. Positive, rational, and real numbers only. Because $\frac{3}{7}$ is a fraction, it is automatically excluded from the groups of natural numbers, whole numbers, and integers. Although it is positive, it cannot be considered even, odd, prime, or composite, because those four classifications apply only to integers.

2. 7. Eliminate double signs to get $6 + 2 - 5 + 4$. You earn a total of 12 ($6 + 2 + 4$), but you lose 5 for a final answer of 7.

3. (a) 40. 8 times 5 equals 40, and because the numbers are both negative (and thus have the same sign), the answer is positive.

    (b) –5. The answer must be negative because the signs are different.

4. $-(8) = -8; |8| = 8$.

5. (a) 10. Subtract $(4 - 2)$ first to get 5×2.

    (b) 2. The innermost grouping symbols, the parentheses, should be done first to get $|2 - 4|$. Subtract $|-2|$, and then take the absolute value: $|-2| = 2$.

6. (a) Commutative property of addition (the order of the numbers differs).

    (b) Additive inverse property (the result is 0, the identity element for addition).

    (c) Associative property of multiplication (the order of the numbers doesn't change, just the way they're grouped).

Chapter 2

1. (a) $\frac{1}{3}$; divide 7 and 21 by the GCF, which is 7.

    (b) $\frac{3}{5}$; GCF = 8.

2. $\frac{10}{30}, \frac{25}{30}, \frac{21}{30}$. The least common denominator is 30. Multiply the numerator and denominator of the first fraction by 10, the second fraction by 5, and the third fraction by 3.

3. $\frac{5}{4}$. Start with common denominators $\left(\frac{18}{12} - \frac{4}{12} + \frac{1}{12}\right)$, combine the numerators $\left(\frac{15}{12}\right)$, and then simplify by dividing the numerator and denominator by 3.

4. 1. Write 8 as a fraction $\left(\frac{5}{6} \times \frac{8}{1} \times \frac{3}{20}\right)$, then multiply numerators together and denominators together to get $\frac{120}{120}$. Any number divides into itself exactly 1 time.

5. $\frac{3}{2}$. Simplify inside the parentheses first, using the least denominator, 14:

    $\frac{7}{14} - \frac{4}{14} = \frac{3}{14}$. That leaves the division problem $\frac{3}{14} \div \frac{1}{7}$, which is equivalent to the multiplication problem $\frac{3}{14} \times \frac{7}{1} = \frac{21}{14}$. Simplify.

Chapter 3

1. $\frac{1}{3}x - 5$. One third of a number equals $\frac{1}{3}x$, and 5 less than that is $\frac{1}{3}x - 5$.

2. 81. $(-3)(-3)(-3)(-3) = 9(-3)(-3) = -27(-3) = 81$.

3. x^9y^{11}. Apply Rules 3 and 4 to get $(x^{15}y^5) \cdot (x^{-6}y^6)$. Then apply Rule 1: $x^{15 + (-6)}y^{5 + 6} = x^9y^{11}$.

4. (a)2.3451×10^{13}.

   (b)1.25×10^{-9}.

5. $12xy^3 - 30y^5 + 48y^3$. Remember, if you're multiplying exponential expressions with the same base, you can add the powers. Otherwise, just list the variables next to one another in alphabetical order.

6. 12. You've got exponents, division, and multiplication, so start with exponents: $100 \div 25 \cdot 3$. Multiplication and division are in the same step, working from left to right: $4 \cdot 3 = 12$.

7. 80. Substitute the values to get $5((-1) - 3)^2$. Work inside parentheses first $(-1 - 3 = -4)$ to get $5(-4)^2$. Because $(-4)^2 = (-4)(-4) = 16$, the answer will be $5 \cdot 16 = 80$.

Chapter 4

1. $x = 11$. To isolate x on the left side, add -8 to both sides of the equation (because -8 is the opposite of 8). That leaves the expression $19 - 8$ on the right side, which should be simplified.

2. $w = -20$. Multiply both sides of the equation by $-\frac{5}{4}$ to get $\frac{20}{20}w = -\frac{80}{4}$, which simplifies to $w = -20$.

3. (a) $x = \frac{17}{6}$. Distribute the 3 to get $6x - 3 = 14$, and add 3 to both sides to separate the variable: $6x = 17$. Divide both sides by 6 (the fraction cannot be simplified).

   (b) $x = -10$. Subtract $4x$ and add 7 on both sides of the equation to get $-2x = 20$. Divide both sides by -2 and simplify.

4. $x = -5$ or $x = 15$. Isolate the absolute value quantity by adding 6 to both sides: $|x - 5| = 10$. Split into two equations ($x - 5 = 10$ and $x - 5 = -10$) and solve them separately.

5. $y = -3x + \dfrac{5}{3}$. Isolate the y-term by subtracting $9x$ from both sides: $3y = -9x + 5$.
    Now divide both sides by 3: $y = \dfrac{-9x+5}{3}$. Write as separate fractions
    $\left(y = \dfrac{-9}{3}x + \dfrac{5}{3} \right)$ and simplify.

Chapter 5

1. $A = (-4,-4)$, $B = (0,3)$, $C = (-2,0)$, $D = (-5,4)$, $E = (2,-1)$, and $F = \left(5\dfrac{1}{2}, 1\dfrac{1}{2} \right)$ or
    $F = \left(\dfrac{11}{2}, \dfrac{3}{2} \right)$.

2. Solve for y to get $y = 4x + 2$. Plugging in $x = -1$, 0, and 1 results in the coordinate
    pairs $(-1,-2)$, $(0,2)$, and $(1,6)$. Plot and connect those points.

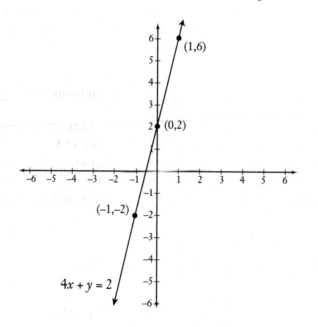

3. Plug in 0 for x to get $2y = -8$, and a y-intercept of $(0,-4)$; plug in 0 for y to get
    $4x = -8$, and an x-intercept of $(-2,0)$. Plot the points and connect them to get the
    graph.

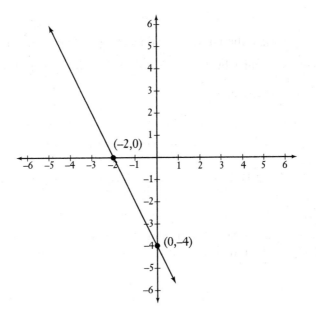

4. $-\dfrac{2}{3}$. Apply the slope formula: $\dfrac{6-(0)}{-5-(4)} = \dfrac{6}{-9}$ and simplify the answer.

5. Set $2x - 4 = 0$ and solve for x to get $x = 2$. Plug this into the equation to get the corresponding y-value that completes the vertex: (2,1). Now choose an x less than 2 and an x greater than 2 (perhaps $x = 0$ and $x = 4$) and calculate the coordinate pairs associated with them. For the x-values indicated, those pairs are (0,5) and (4,5). Draw two lines that begin at the vertex and pass through those points.

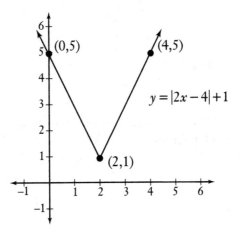

Chapter 6

1. $y = 4x - 15$. Substitute $m = 4$, $x_1 = 2$, and $y_1 = -7$ into the point-slope formula to get $y - (-7) = 4(x - 2)$. Simplify both sides ($y + 7 = 4x - 8$) and finish by subtracting 7 from both sides.

2. Slope $= -\dfrac{3}{2}$; y-intercept = (0,2). Subtract $3x$ from both sides and divide by 2 to solve for y: $y = -\dfrac{3}{2}x + 2$. The slope is the coefficient of the x-term, and the y-intercept is the constant.

3. Solve for y to get $y = -2x - 1$. Note that slope *must* be a fraction, but because -2 is an integer, it is also rational: $-\dfrac{2}{1}$. Either count down two units and right one from the y-intercept of (0,–1), or count up two units and left one to get another point on the line.

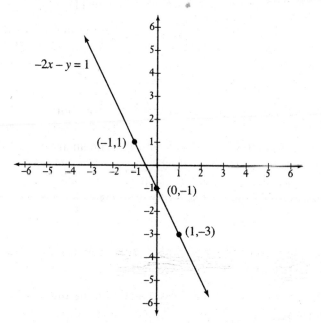

4. $2x - 15y = -28$. Multiply everything by the least common denominator of 12 to eliminate fractions: $15y = 28 + 2x$. Subtract $2x$ from both sides of the equation: $-2x + 15y = 28$. Finally, multiply everything by -1 to make the x-term positive.

5. $x - y = -8$. First calculate the slope: $m = \dfrac{0 - (5)}{-8 - (-3)} = \dfrac{-5}{-5} = 1$. Now apply the point-slope formula using one of the points. If you choose the point (–8,0), you get $y = 1(x - (-8))$, or $y = x + 8$. Put this equation in standard form.

6. $x - 3y = -9$. Put $-2x + 6y = 7$ in slope-intercept form $\left(y = \dfrac{1}{3}x + \dfrac{7}{6} \right)$ to determine its slope $\left(\dfrac{1}{3} \right)$. Line k must have the same slope, so apply the point-slope formula using the given point $(-6,1)$ to get $y - (1) = \dfrac{1}{3}(x - (-6))$, which simplifies to $y - 1 = \dfrac{1}{3}x + 2$. Put that equation in standard form.

Chapter 7

1. (a) False; no number can be less than itself. Had the statement been $-3 \le -3$, however, it would have been true.

   (b) True; 5 is either less than *or* equal to 11 (only one of those two conditions can be true at a time, and it's true that 5 is less than 11).

2. $w \le 15$. First, simplify the left side by distributing the 2: $2w - 12 \le 18$. Isolate the variable by adding 12 to both sides: $2w \le 30$. Divide both sides by 2 to get the final answer. No need to reverse the inequality sign, as you're dividing both sides by a positive number.

3. Solve the inequality by subtracting $2x$ and 1 from both sides to get $-3x \ge -6$. Now divide both sides by -3 to get $x \le 2$. (Don't forget to reverse the inequality sign.) The graph consists of a solid dot at 2 and an arrow extending from there to the left.

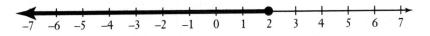

4. $-3 < x < 4$. Subtract 5 from all three parts to get $-6 < 2x < 8$, and then divide everything by 2 to get the final answer.

5. Place a solid dot at -2 and an open dot at 5 on the number line, and connect the dots with a dark line.

6. $3 < x < 7$. Divide both sides by 4 to isolate the absolute value quantity: $|x - 5| < 2$. Rewrite the statement as $-2 < x - 5 < 2$ and solve by adding 5 to everything. The graph is a line segment with open endpoints at 3 and 7.

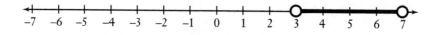

7. $x \le 2$ or $x \ge 6$. Split the inequality appropriately ($x - 4 \ge 2$ or $x - 4 \le -2$) and solve. Both inequality symbols specify "or equal to," so make sure to use solid dots on the graph.

8. Notice that the inequality is in slope-intercept form with slope $m = -2$ and y-intercept $(0,3)$. The inequality symbol specifies "or equal to," so the line should be solid. If you test the point $(0,0)$, you get $0 \ge 3$, which is *false*, so you should shade the region *not* containing the origin.

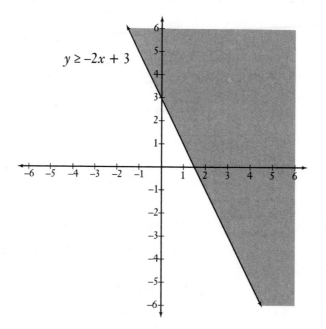

Chapter 8

1. (0,3). Rewrite both equations in slope-intercept form.

$$\begin{cases} y = -2x + 3 \\ y = \dfrac{1}{3}x + 3 \end{cases}$$

Notice that both graphs have the same y-intercept.

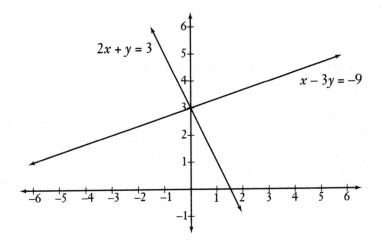

2. (3,–2). The first equation has an x-coefficient of 1, so solve it for x to get $x = 4y + 11$. Plug that into the other equation.

$$3(4y+11)+7y=-5$$
$$12y+33+7y=-5$$
$$19y+33=-5$$
$$19y=-38$$
$$y=-2$$

Substitute $y = -2$ into $x = 4y + 11$ to get the corresponding x-value: $x = 3$.

3. (–2,7). Multiply the first equation by 2 to get $4x - 2y = -22$ and add it to the second equation.

$$
\begin{array}{rcrcr}
4x & - & 2y & = & -22 \\
5x & + & 2y & = & 4 \\
\hline
9x & & & = & -18
\end{array}
$$

Substitute $x = -2$ into one of the original equations to get the final answer.

$$5x + 2y = 4$$
$$5(-2) + 2y = 4$$
$$-10 + 2y = 4$$
$$2y = 14$$
$$y = 7$$

4. Dependent. Multiply the first equation by 2 and add it to the second equation. Every term cancels out.

$$
\begin{array}{rrrrr}
8x & + & 6y & = & -4 \\
-8x & - & 6y & = & 4 \\
\hline
0 & + & 0 & = & 0
\end{array}
$$

Because the variable-free statement $0 = 0$ is true, the system is dependent.

5. Graph the inequalities, which are already in slope-intercept form. You'll end up shading left of the dotted line and below the solid line.

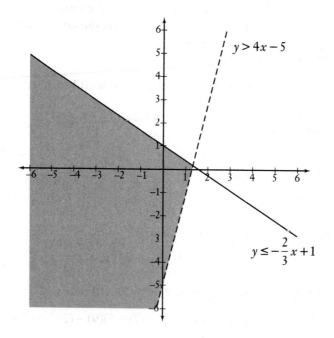

$y > 4x - 5$

$y \le -\dfrac{2}{3}x + 1$

Chapter 9

1. $\begin{bmatrix} 4 & 16 & 36 \\ -8 & 28 & -12 \\ 44 & -20 & 24 \end{bmatrix}$. Multiply each element of B by 4 to get $4B$.

2. $\begin{bmatrix} -7 & 14 \\ 41 & -6 \end{bmatrix}$. Multiply A by the scalar 4 and B by the scalar –3.

$$4A-3B = \begin{bmatrix} 4(-1) & 4(1/2) \\ 4(5) & 4(3) \end{bmatrix} + \begin{bmatrix} -3(1) & -3(-4) \\ -3(-7) & -3(6) \end{bmatrix}$$

$$4A-3B = \begin{bmatrix} -4 & 2 \\ 20 & 12 \end{bmatrix} + \begin{bmatrix} -3 & 12 \\ 21 & -18 \end{bmatrix}$$

$$4A-3B = \begin{bmatrix} -4-3 & 2+12 \\ 20+21 & 12-18 \end{bmatrix}$$

3. The product $B \cdot A$ exists because the number of columns in B equals the number of rows in A. The order of $B \cdot A$ is 3 × 4; it has the same number of rows as B and the same number of columns as A.

4. $\begin{bmatrix} 17 & -19 \\ -19 & 9 \\ -24 & 8 \end{bmatrix}$. The product matrix has three rows (like the left matrix) and two columns (like the right matrix).

$$\begin{bmatrix} 2(6)+(-5)(-1) & 2(-2)+(-5)(3) \\ (-3)(6)+1(-1) & (-3)(-2)+1(3) \\ -4(6)+0(-1) & -4(-2)+0(3) \end{bmatrix} = \begin{bmatrix} 12+5 & -4-15 \\ -18-1 & 6+3 \\ -24+0 & 8+0 \end{bmatrix}$$

5. 3. Multiply 9 times –1 and then subtract 3 times –4.

$$9(-1) - 3(-4) = -9 + 12 = 3$$

6. 279. Multiply along the diagonals illustrated in Figure 9.3.

$$(-2)(-5)(-3)+(-4)(1)(7)+(9)(3)(2)-(7)(-5)(9)-(2)(1)(-2)-(-3)(3)(-4)$$
$$=-30-28+54+315+4-36$$
$$=279$$

7. $\left(-\frac{1}{3}, 1\right)$. The Cramer's Rule matrices are $C = \begin{bmatrix} 6 & -5 \\ 3 & 2 \end{bmatrix}$, $A = \begin{bmatrix} -7 & -5 \\ 1 & 2 \end{bmatrix}$, and $B = \begin{bmatrix} 6 & -7 \\ 3 & 1 \end{bmatrix}$. The determinants are $|C| = 12 - (-15) = 27$, $|A| = -14 - (-5) = -9$, and $|B| = 6 - (-21) = 27$. Therefore, $x = \frac{|A|}{|C|} = \frac{-9}{27} = -\frac{1}{3}$ and $y = \frac{|B|}{|C|} = \frac{27}{27} = 1$.

Chapter 10

1. (a) Because it has two terms and a degree of 3, $4x^3 + 2$ is a cubic binomial.

   (b) If x is not raised to an explicit power, then the power is understood to be 1. So, the polynomial can be rewritten as $11x^1$. It has one term with a variable raised to a power of 1, so $11x$ is a linear monomial.

2. $5x^2 - 12x + 1$. Start by distributing the 2 to get $3x^2 - 9 + 2x^2 - 12x + 10$. In this polynomial, $3x^2$ and $2x^2$ are like terms, as are -9 and 10, so combine them. The $-12x$ term has no like terms in the polynomial, so leave it alone.

3. $15x^5y + 12x^4y^2 - 6x^2y^6$. Distribute $3x^2y$ to each term and calculate the products: $(3x^2y)(5x^3) = 15x^5y$, $(3x^2y)(4x^2y) = 12x^4y^2$, and $(3x^2y)(-2y^5) = -6x^2y^6$. Write the terms in decreasing order of the exponents of x (because x comes before y alphabetically).

4. $2x^2 - 5xy - 3y^2$. Distribute $2x$ and then y to the terms of the expression $x - 3y$.

$$2x(x) + 2x(-3y) + y(x) + y(-3y) = 2x^2 - 6xy + xy - 3y^2$$

   Combine like terms $-6xy$ and xy: $-6xy + xy = -5xy$.

5. $x - 11 + \frac{52}{x+4}$. Perform long division.

$$
\begin{array}{r}
x - 11 \\
x+4{\overline{\smash{\big)}\,x^2 - 7x + 8}} \\
\underline{-x^2 - 4x} \\
-11x + 8 \\
\underline{11x + 44} \\
52
\end{array}
$$

6. $4x^2 + 6x + 2 + \frac{5}{x-2}$. Perform synthetic division.

$$
\begin{array}{r|rrrr}
2 & 4 & -2 & -10 & 1 \\
  &   & 8 & 12 & 4 \\
\hline
  & 4 & 6 & 2   & 5
\end{array}
$$

Chapter 11

1. $3x^3y^2(3x^2 + xy - 2y^5)$. The GCF of the coefficients is 3. The lowest power of x is x^3, and the lowest power of y is y^2, so the GCF of the polynomial is $3x^3y^2$. Divide the terms of the original polynomial by the GCF.

$$\frac{9x^5y^2}{3x^3y^2} + \frac{3x^4y^3}{3x^3y^2} - \frac{6x^3y^7}{3x^3y^2}$$

$$= \frac{9}{3}x^{5-3}y^{2-2} + \frac{3}{3}x^{4-3}y^{3-2} - \frac{6}{3}x^{3-3}y^{7-2}$$

$$= 3x^2y^0 + 1x^1y^1 - 2x^0y^5$$

$$= 3x^2 + xy - 2y^5$$

Multiply this new expression by the GCF $3x^3y^2$ to finish.

2. $(2x + 1)(6x^3 + 7)$. The first two terms have a GCF of $6x^3$ and the others have a GCF of 7. Factor out the GCFs to get $6x^3(2x + 1) + 7(2x + 1)$. Finally, factor out the common binomial $(2x + 1)$ and write what's left in parentheses.

3. $5(x + 5)(x - 5)$. Start by factoring out the GCF 5 to get $5(x^2 - 25)$. Because $x^2 - 25$ is the difference of perfect squares ($a^2 - b^2$ when $a = x$ and $b = 5$), it must be factored further: $x^2 - 25 = (x + 5)(x - 5)$.

4. $2x(x - 8)(x - 4)$. Factor out the GCF $2x$: $2x(x^2 - 12x + 32)$. Because 32 is positive, the signs of the mystery numbers match, and because they add up to a negative number (-12), both numbers must be negative. The only two negative numbers that add up to -12 and multiply to 32 are -8 and -4.

5. $(4x - 1)(x + 6)$. The two numbers that add up to 23 and have a product of $4(-6) = -24$ are 24 and -1. Rewrite the polynomial as $4x^2 + (24 - 1)x - 6$ and distribute x to get $4x^2 + 24x - x - 6$; factor by grouping to finish.

$$4x(x + 6) - 1(x + 6) = (x + 6)(4x - 1)$$

Chapter 12

1. $10|x^3y|\sqrt{3y}$. The radical's a square root, so only perfect squares escape. Rewrite the coefficient as $100 \cdot 3$, or $10^2 \cdot 3$. Notice that x^6 is a perfect square—it can be written $x^3 \cdot x^3$ or $(x^3)^2$, so it's raised to the second power. The best you can do for y^3 is to rewrite it as $y^2 \cdot y$, so it contains one perfect square: $\sqrt{10^2 \cdot 3 \cdot (x^3)^2 \cdot y^2 \cdot y}$.

Both x^3 and y are raised to a power that matches the even index of the radical, so they must be surrounded by absolute value bars.

2. 125. Rewrite $25^{3/2}$ as $\left(\sqrt{25}\right)^3$. Because $\sqrt{25} = \sqrt{5^2} = 5$, that means $\left(\sqrt{25}\right)^3 = 5^3 = 125$.

3. $6x\sqrt[3]{x}$. Rewrite the first radical as $\sqrt[3]{2^3 \cdot x^3 \cdot x}$, so that it contains as many perfect cubes as possible. Once those perfect cubes are paroled, the radical simplifies to $2x\sqrt[3]{x}$, and the original problem becomes $2x\sqrt[3]{x} + 4x\sqrt[3]{x}$. Now you have like radicals with coefficients that are like terms, so add the coefficients together: $(2x + 4x)\sqrt[3]{x} = 6x\sqrt[3]{x}$.

4. $6|xy|\sqrt{x}$. Multiply the radicands to get $\sqrt{36x^3 y^2}$ and simplify the radical:

$\sqrt{6^2 \cdot x^2 \cdot x \cdot y^2} = 6|xy|\sqrt{x}$. Don't forget both x and y need to be in absolute value signs, they are raised to an even power that matches the index.

5. $\dfrac{\sqrt{xy}}{3|x|}$. Simplify the fraction: $\sqrt{\dfrac{2x^2 y^3}{18x^3 y^2}}$.

$$\sqrt{\frac{2}{18} x^{2-3} y^{3-2}} = \sqrt{\frac{1}{9} x^{-1} y^1} = \sqrt{\frac{y}{9x}} = \frac{\sqrt{y}}{\sqrt{3^2 \cdot x}} = \frac{\sqrt{y}}{3\sqrt{x}}$$

Rationalize by multiplying the numerator and denominator by $\sqrt{x}$.

$$\frac{\sqrt{y}}{3\sqrt{x}}\left(\frac{\sqrt{x}}{\sqrt{x}}\right) = \frac{\sqrt{xy}}{3\sqrt{x^2}} = \frac{\sqrt{xy}}{3|x|}$$

6. $x = 19$. Divide both sides by 2 to isolate the radical: $\sqrt{x-3} = 4$. The index of the radical is 2, so raise both sides to the second power.

$$\left(\sqrt{x-3}\right)^2 = 4^2$$
$$x - 3 = 16$$

Solve the equation by adding 3 to both sides.

7. $5i$. Rewrite the left term using a perfect square and rewrite the right term using powers of i^2.

$$\sqrt{-36} + i^3 = \sqrt{-1 \cdot 6^2} + i^{2+1}$$
$$= \sqrt{-1} \cdot \sqrt{6^2} + i^2 \cdot i$$
$$= i \cdot 6 + (-1)i$$
$$= 6i - i$$
$$= -5i$$

8. (a) $(3 - 4i) + (8 + i) = (3 + 8) + (-4i + i) = 11 - 3i$

(b) $(3 - 4i) - (8 + i) = (3 - 4i) + (-8 - i) = (3 - 8) + (-4i - i) = -5 - 5i$

(c) $(3 - 4i)(8 + i) = 3(8) + 3(i) + (-4i)(8) + (-4i)(i)$

$$= 24 + 3i - 32i - 4i^2$$
$$= 24 - 29i - 4(-1)$$
$$= 28 - 29i$$

(d) $\dfrac{3 - 4i}{8 + i}\left(\dfrac{8 - i}{8 - i}\right) = \dfrac{3(8) + 3(-i) + (-4i)(8) + (-4i)(-i)}{8(8) + 8(-i) + i(8) + i(-i)}$

$$= \dfrac{24 - 3i - 32i + 4i^2}{64 - 8i + 8i - i^2}$$
$$= \dfrac{24 - 35i + 4(-1)}{64 - (-1)}$$
$$= \dfrac{20 - 35i}{65}$$
$$= \dfrac{20}{65} - \dfrac{35}{65}i$$
$$= \dfrac{4}{13} - \dfrac{7}{13}i$$

Chapter 13

1. $x = -\dfrac{5}{2}, 0, \dfrac{5}{2}$. Subtract $25x$ from both sides of the equation and factor out the GCF to get $x(4x^2 - 25) = 0$. The quantity in parentheses is a difference of perfect squares, so factor it: $x(2x + 5)(2x - 5) = 0$. Now set each factor equal to 0 and solve.

$$2x + 5 = 0 \qquad\qquad 2x - 5 = 0$$
$$x = 0 \quad \text{or} \qquad 2x = -5 \quad \text{or} \qquad 2x = 5$$
$$x = -\dfrac{5}{2} \qquad\qquad x = \dfrac{5}{2}$$

2. $x = -3 - 2\sqrt{3}$ or $x = -3 + 2\sqrt{3}$. The coefficient of x^2 is already 1, so add 3 to both sides of the equation: $x^2 + 6x = 3$. According to the bug, you should add the square of half of 6, or $3^2 = 9$, to both sides.

$$x^2 + 6x + 9 = 3 + 9$$
$$x^2 + 6x + 9 = 12$$

Factor the perfect square and take the square root of both sides of the equation to solve for x.

$$(x+3)^2 = 12$$

$$\sqrt{(x+3)^2} = \pm\sqrt{12}$$

$$x+3 = \pm 2\sqrt{3}$$

$$x = -3 \pm 2\sqrt{3}$$

3. $x = -3 - 2\sqrt{3}$ or $x = -3 + 2\sqrt{3}$. In this problem, $a = 1$, $b = 6$, and $c = -3$.

$$x = \frac{-6 \pm \sqrt{36 - 4(1)(-3)}}{2(1)} = \frac{-6 \pm \sqrt{48}}{2} = -\frac{6}{2} \pm \frac{\sqrt{4^2 \cdot 3}}{2} = -3 \pm \frac{4\sqrt{3}}{2} = -3 \pm 2\sqrt{3}$$

4. 1. Set $a = 25$, $b = -40$, and $c = 16$ and evaluate the discriminant.

$$b^2 - 4ac = (-40)^2 - 4(25)(16)$$

$$= 1600 - 1600$$

$$= 0$$

A discriminant of 0 indicates that the quadratic has one real solution, a double root.

5. $x \leq -3$ or $x \geq \frac{1}{2}$. Factor the quadratic to get $(2x-1)(x+3) \geq 0$, which yields the critical numbers $x = \frac{1}{2}$ and $x = -3$. They split the number line into the intervals $x \leq -3$, $-3 \leq x \leq \frac{1}{2}$, and $x \geq \frac{1}{2}$. Choose one test value from each interval (like $x = -4$, $x = 0$, $x = 1$) and substitute each into the inequality.

Test $x = -4$	Test $x = 0$	Test $x = 1$
$2(-4)^2 + 5(-4) - 3 \geq 0$	$2(0)^2 + 5(0) - 3 \geq 0$	$2(1)^2 + 5(1) - 3 \geq 0$
$2(16) - 20 - 3 \geq 0$	$0 - 3 \geq 0$	$2 + 5 - 3 \geq 0$
$9 \geq 0$	$-3 \geq 0$	$4 \geq 0$
True	False	True

The intervals $x \leq -3$ and $x \geq \frac{1}{2}$ together comprise the solution, so darken those intervals on a number line (marking the critical numbers with solid dots because of the inequality symbol $\geq$) to complete the graph.

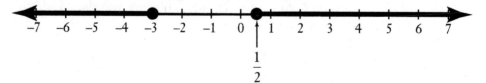

Chapter 14

1. $(x + 2)(x - 2)(2x + 5)$. Because $x + 2$ is linear, you can apply synthetic division.

$$
\begin{array}{r|rrrr}
-2 & 2 & 5 & -8 & -20 \\
   &   & -4 & -2 & 20 \\
\hline
   & 2 & 1 & -10 & 0
\end{array}
$$

Therefore, $2x^3 + 5x^2 - 8x - 20 = (x + 2)(2x^2 + x - 10)$. Use the bomb method to factor the quadratic: $2x^2 + x - 10 = (2x + 5)(x - 2)$.

2. -9 and 1. Apply synthetic division to determine whether -2 is a root.

$$
\begin{array}{r|rrrr}
-2 & 1 & 10 & 7 & -18 \\
   &   & -2 & -16 & 18 \\
\hline
   & 1 & 8 & -9 & 0
\end{array}
$$

Because -2 is a root, $(x - (-2)) = x + 2$ is a factor of the polynomial.

$$x^3 + 10x^2 + 7x - 18 = 0$$
$$(x + 2)(x^2 + 8x - 9) = 0$$

Factor the quadratic and solve.

$$(x + 2)(x + 9)(x - 1) = 0$$

$$
\begin{array}{ccc}
x + 2 = 0 & x + 9 = 0 & x - 1 = 0 \\
x = -2 \quad \text{or} & x = -9 \quad \text{or} & x = 1
\end{array}
$$

3. $-2, -\dfrac{1}{2}, \dfrac{1}{2}$, and 1. Apply synthetic division to demonstrate that $x = 1$ is a root.

$$
\begin{array}{r|rrrrr}
1 & 4 & 4 & -9 & -1 & 2 \\
  &   & 4 & 8 & -1 & -2 \\
\hline
  & 4 & 8 & -1 & -2 & 0
\end{array}
$$

Similarly, -2 is a root of that quotient, $4x^3 + 8x^2 - x - 2$.

$$
\begin{array}{r|rrrr}
-2 & 4 & 8 & -1 & -2 \\
   &   & -8 & 0 & 2 \\
\hline
   & 4 & 0 & -1 & 0
\end{array}
$$

Therefore, $4x^4 + 4x^3 - 9x^2 - x + 2 = (x - 1)(x + 2)(4x^2 - 1)$. The quadratic factor is a difference of perfect squares: $4x^2 - 1 = (2x + 1)(2x - 1)$. Rewrite the original equation in factored form and solve.

$$4x^4 + 4x^3 - 9x^2 - x + 2 = 0$$
$$(x-1)(x+2)(2x+1)(2x-1) = 0$$

$$x - 1 = 0 \quad \text{or} \quad x + 2 = 0 \quad \text{or} \quad 2x + 1 = 0 \quad \text{or} \quad 2x - 1 = 0$$
$$x = 1 \qquad\qquad x = -2 \qquad\qquad x = -\frac{1}{2} \qquad\qquad x = \frac{1}{2}$$

4. $3, -\dfrac{1}{4} - \dfrac{i\sqrt{55}}{4}$, and $-\dfrac{1}{4} + \dfrac{i\sqrt{55}}{4}$. The only rational root is 3.

$$
\begin{array}{r|rrrr}
3 & 2 & -5 & 4 & -21 \\
  &   & 6 & 3 & 21 \\
\hline
  & 2 & 1 & 7 & 0
\end{array}
$$

Factor the polynomial to get $(x - 3)(2x^2 + x + 7) = 0$. Set both factors equal to 0 and solve.

$$x - 3 = 0 \quad \text{or} \quad 2x^2 + x + 7 = 0$$

The solution to the left equation is $x = 3$. Use the quadratic formula to solve the equation on the right.

$$x = \frac{-1 \pm \sqrt{1^2 - 4(2)(7)}}{2(2)} = \frac{-1 \pm \sqrt{-55}}{4} = -\frac{1}{4} \pm \frac{\sqrt{-55}}{4} = -\frac{1}{4} \pm \frac{i\sqrt{55}}{4}$$

Chapter 15

1.   (a) –18. Evaluate $h(-1)$ and $k(-1)$.

$$
\begin{aligned}
h(x) &= x^2 + 7x - 5 & k(x) &= 4x - 3 \\
h(-1) &= (-1)^2 + 7(-1) - 5 & k(-1) &= 4(-1) - 3 \\
h(-1) &= 1 - 7 - 5 & k(-1) &= -4 - 3 \\
h(-1) &= -11 & k(-1) &= -7
\end{aligned}
$$

Sum the results.

$$(h + k)(-1) = h(-1) + k(-1) = -11 + (-7) = -18$$

(b) –4. Use the values for $h(-1)$ and $k(-1)$ calculated in part (a).

$$(h - k)(-1) = h(-1) - k(-1) = -11 - (-7) = -11 + 7 = -4$$

(c) 77. Use the values for $h(-1)$ and $k(-1)$ calculated in part (a).

$$(hk)(-1) = h(-1) \cdot k(-1) = (-11)(-7) = 77$$

(d) $\dfrac{11}{7}$. Use the values for $h(-1)$ and $k(-1)$ calculated in part (a).

$$\left(\frac{h}{k}\right)(-1) = \frac{h(-1)}{k(-1)} = \frac{-11}{-7} = \frac{11}{7}$$

2. (a) $f\left(\sqrt{x-2}\right) = x + 3$. Substitute $g(x) = \sqrt{x-2}$ for x in the function $f(x) = x^2 + 5$.

$$f\left(\sqrt{x-2}\right) = \left(\sqrt{x-2}\right)^2 + 5$$
$$= (x-2) + 5$$
$$= x + 3$$

(b) $g\left(x^2 + 5\right) = \sqrt{x^2 + 3}$. This time, substitute $f(x)$ into $g(x)$.

$$g\left(x^2 + 5\right) = \sqrt{\left(x^2 + 5\right) - 2}$$
$$= \sqrt{x^2 + 3}$$

Note that the expression $\sqrt{x^2 + 3}$ cannot be simplified. While it's true that x^2 is a perfect square, you can release it from the radical prison only if it's *multiplied* by the other contents of the radicand, not added to them.

3. If $f(x)$ and $g(x)$ are inverses, then $f(g(x)) = g(f(x)) = x$.

$$f(g(x)) = \frac{3}{1}\left(\frac{x+5}{3}\right) - 5$$
$$= \frac{3x + 15}{3} - 5 \qquad\qquad g(f(x)) = \frac{(3x - 5) + 5}{3}$$
$$= \frac{3x}{3} + \frac{15}{3} - 5 \qquad\qquad\qquad = \frac{3x}{3}$$
$$= x + 5 - 5 \qquad\qquad\qquad\qquad = x$$
$$= x$$

4. $g^{-1}(x) = \dfrac{1}{7}x + \dfrac{3}{7}$. Rewrite $g(x)$ as y and reverse the x and y in the equation to get $x = 7y - 3$. Solving for y gives you $y = \dfrac{x+3}{7}$ or $y = \dfrac{1}{7}x + \dfrac{3}{7}$. Replace y with $g^{-1}(x)$.

5. $f(1) < f(-1) < f(0)$. Because $x = -1$ and $x = 0$ satisfy the condition $x \leq 0$, substitute them into the first expression $(3x + 4)$; $x = 1$ should be plugged into $x - x^2$, as $3 > 0$.

$$f(-1) = 3(-1) + 4 \qquad f(0) = 3(0) + 4 \qquad f(1) = 1 - (1)^2$$
$$= -3 + 4 \qquad\qquad = 4 \qquad\qquad = 1 - 1$$
$$= 1 \qquad\qquad\qquad\qquad\qquad\qquad = 0$$

Chapter 16

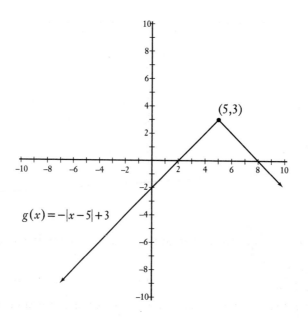

$$g(x) = -|x - 5| + 3$$

(5,3)

1. As discussed in Chapter 5, the graph is v-shaped.

2. According to the vertical line test, $g(x)$ is a function; however, $g(x)$ fails the horizontal line test; any horizontal line $y = c$ will pass through the graph twice if $c < 3$, so $g(x)$ isn't one-to-one.

3. The graph extends infinitely to the right and left, so any vertical line will intersect it. Therefore, the domain of $g(x)$ is all real numbers (nothing is disqualified). However, only horizontal lines drawn at a height of 3 or below will hit the graph, so the range is $y \leq 3$.

4. Because you're plugging in $-x$ instead of x, reflect the graph of x^3 about the y-axis. Next, move the graph down two units, as 2 is subtracted from x^3.

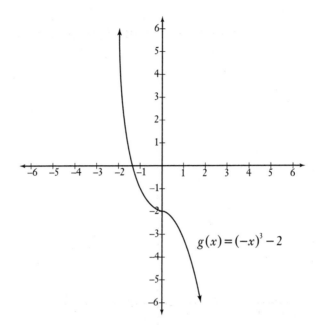

$$g(x) = (-x)^3 - 2$$

Chapter 17

1. $\dfrac{x^2 + 3x}{3x + 5}$. Factor the numerator and denominator completely to get

    $\dfrac{x(x+3)(2x-1)}{(2x-1)(3x+5)}$. Eliminate the common factor $2x - 1$ to get $\dfrac{x(x+3)}{3x+5}$. You can leave your answer like that or multiply the terms in the numerator.

2. $\dfrac{x^2 + 3x + 16}{(x-2)(x+2)(x-5)}$. Factor the denominators to get

    $\dfrac{x}{(x+2)(x-2)} + \dfrac{8}{(x-5)(x-2)}$. The LCD is $(x+2)(x-2)(x-5)$. Multiply the top and bottom of each fraction by the factors needed to reach the least common denominator.

    $$\left(\frac{x-5}{x-5}\right) \cdot \frac{x}{(x+2)(x-2)} + \frac{8}{(x-5)(x-2)} \cdot \left(\frac{x+2}{x+2}\right)$$

    $$= \frac{x^2 - 5x + 8x + 16}{(x-2)(x+2)(x-5)}$$

3. $\dfrac{x+2}{(x-4)(x+3)}$ or $\dfrac{x+2}{x^2-x-12}$. Take the reciprocal of the second fraction, factor everything, and multiply the fractions together.

$$\frac{x^2-4x-12}{3x^2-10x-8}\cdot\frac{3x+2}{x^2-3x-18}=\frac{(x-6)(x+2)(3x+2)}{(3x+2)(x-4)(x+3)(x-6)}$$

Simplify the fraction.

$$\frac{\cancel{(x-6)}\,(x+2)\,\cancel{(3x+2)}}{\cancel{(3x+2)}\,(x-4)(x+3)\,\cancel{(x-6)}}=\frac{x+2}{(x-4)(x+3)}$$

4. $(x-4)(x+2)$ or x^2-2x-8. Rewrite the complex fraction as a division problem, and then use the reciprocal to rewrite it as a multiplication problem.

$$\frac{x^2-7x+12}{x+2}\div\frac{x-3}{x^2+4x+4}=\frac{\left(x^2-7x+12\right)\left(x^2+4x+4\right)}{(x+2)(x-3)}$$

Factor and simplify.

$$\frac{\cancel{(x-3)}\,(x-4)\,\cancel{(x+2)}\,(x+2)}{\cancel{(x+2)}\,\cancel{(x-3)}}=\frac{(x-4)(x+2)}{1}$$

Because the denominator is 1, there's no need to include it in your answer.

Chapter 18

1. $x=-\dfrac{11}{6}$. Factor the second denominator into $(x-8)(x+2)$ and then multiply the entire equation by the least common denominator, $(x-8)(x+2)$.

$$\frac{(x-8)(x+2)}{1}\cdot\left[\frac{x+3}{x-8}+\frac{x}{(x-8)(x+2)}\right]=\frac{(x-8)(x+2)}{1}\cdot(1)$$

$$\frac{\cancel{(x-8)}\,(x+2)(x+3)}{\cancel{x-8}}+\frac{x\,\cancel{(x-8)}\,\cancel{(x+2)}}{\cancel{(x-8)}\,\cancel{(x+2)}}=\frac{(x-8)(x+2)}{1}$$

$$(x+2)(x+3)+x=(x-8)(x+2)$$

$$x^2+5x+6+x=x^2-6x-16$$

$$x^2+6x+6=x^2-6x-16$$

Solve for x.

$$x^2 - x^2 + 6x + 6x = -16 - 6$$
$$12x = -22$$
$$x = -\frac{22}{12}$$
$$x = -\frac{11}{6}$$

2. $x = -2$ or 1. Cross multiply to change the proportion into a quadratic equation and solve.

$$(3x - 2)(x + 1) = x(2x)$$
$$3x^2 + 3x - 2x - 2 = 2x^2$$
$$x^2 + x - 2 = 0$$
$$(x + 2)(x - 1) = 0$$
$$x + 2 = 0 \qquad x - 1 = 0$$
$$\qquad\qquad \text{or}$$
$$x = -2 \qquad\qquad x = 1$$

3. $y = \dfrac{175}{4}$ (or 43.75). Find k, the constant of proportionality: $k = \dfrac{y}{x} = \dfrac{15}{12} = \dfrac{5}{4}$. Now use the equation $k = \dfrac{y}{x}$ again, this time substituting $k = \dfrac{5}{4}$ and $x = 35$: $\dfrac{5}{4} = \dfrac{y}{35}$. Cross multiply and solve.

$$5(35) = 4 \cdot y$$
$$175 = 4y$$
$$\frac{175}{4} = y$$

4. $x = \dfrac{675}{2}$ (or 337.5). Because x and y vary inversely, $xy = k$.

$$9(75) = k$$
$$675 = k$$

Now solve the equation $xy = k$ for x when $y = 2$ and $k = 675$.

$$xy = k$$
$$x(2) = 675$$
$$x = \frac{675}{2}$$

5. $x < 2$ or $x > \dfrac{13}{2}$. Subtract 3 from both sides to get zero on the right side of the inequality, and combine the terms into one fraction.

$$\frac{x+7}{x-2} - 3 < 0$$

$$\frac{x+7}{x-2} - \left(\frac{x-2}{x-2}\right)\cdot\frac{3}{1} < 0$$

$$\frac{x+7}{x-2} - \frac{3x-6}{x-2} < 0$$

$$\frac{x+7-(3x-6)}{x-2} < 0$$

$$\frac{x+7-3x+6}{x-2} < 0$$

$$\frac{-2x+13}{x-2} < 0$$

Set the numerator and denominator equal to zero and solve to identify the critical numbers; plot them both on the number line using open dots.

$$-2x+13 = 0$$
$$-2x = -13 \qquad \text{or} \qquad \begin{array}{c} x-2 = 0 \\ x = 2 \end{array}$$
$$x = \frac{13}{2} = 6.5$$

The critical numbers split the number line into three intervals: $x < 2$, $2 < x < \dfrac{13}{2}$, and $x > \dfrac{13}{2}$. Choose test values from each interval (like $x = 0$, $x = 4$, and $x = 10$) and substitute them into the inequality.

$\boxed{\text{Test } x = 0}$	$\boxed{\text{Test } x = 4}$	$\boxed{\text{Test } x = 10}$
$\dfrac{-2x+13}{x-2} < 0$	$\dfrac{-2x+13}{x-2} < 0$	$\dfrac{-2x+13}{x-2} < 0$
$\dfrac{-2(0)+13}{0-2} < 0$	$\dfrac{-2(4)+13}{4-2} < 0$	$\dfrac{-2(10)+13}{10-2} < 0$
$-\dfrac{13}{2} < 0$	$\dfrac{5}{2} < 0$	$-\dfrac{7}{8} < 0$
$\boxed{\text{True}}$	$\boxed{\text{False}}$	$\boxed{\text{True}}$

Because the intervals $x < 2$ and $x > \dfrac{13}{2}$ result in true statements, they comprise the solution. Darken the intervals on the number line to graph.

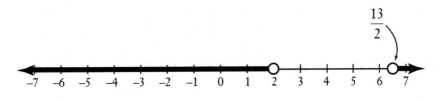

Chapter 19

1. (a)$13,250. Plug $p = 5,000$, $r = 0.0825$, and $t = 20$ into the simple interest formula.

$$i = prt$$
$$= (5,000)(0.0825)(20)$$
$$= 8,250$$

Add the interest to the principal to calculate the balance: $8,250 + $5,000 = $13,250.

(b)$25,999.84. Apply the compound interest formula with $p = 5,000$, $r = 0.0825$, $t = 20$, and $n = 52$.

$$b = p\left(1 + \frac{r}{n}\right)^{nt}$$
$$= 5,000(1.0015865)^{(52)(20)}$$
$$= 5,000(5.1999684)$$
$$= 25,999.84$$

2. $r = \sqrt[3]{18}$ inches. Because the height is twice the radius, $h = 2r$. You're given the volume of the cylinder, so use the corresponding formula, $V = \pi r^2 h$, plug in the values you know, and solve for r.

$$36\pi = \pi r^2 (2r)$$
$$36\pi = 2\pi r^3$$
$$\frac{36\pi}{2\pi} = r^3$$
$$18 = r^3$$
$$\sqrt[3]{18} = r$$

3. 7.08 hours. Think of the trip to the store as trip A and the trip home as trip B. Use two distance formulas, $D_A = r_A \cdot t_A$ and $D_B = r_B \cdot t_B$. Set $r_A = 17$ and $t_A = 1.25$, so $D_A = 17(1.25) = 21.25$, meaning that the distance between home and the store is 21.25 miles. During the trip home, his speed is 3 mph, so $r_B = 3$. The distance home is the same as the distance to the store, so $D_B = D_A = 21.25$. Apply the formula $D_B = r_B \cdot t_B$ and solve for t_B.

$$D_B = r_B \cdot t_B$$
$$21.25 = 3(t_B)$$
$$\frac{21.25}{3} = t_B$$
$$7.08 \approx t_B$$

4. 3.87%. Ingredient 1 is the 7 gallons of water with a descriptive value of 1.2% = 0.012. Ingredient 2 is the 3 gallons of water that will fill the tank, but you don't know its saline value, so use the variable x to represent it. The total volume of the tank is 10 gallons. That number should be multiplied by 2% = 0.02. Create the mixture equation and solve for x.

$$7(0.012) + 3x = 10(0.02)$$
$$0.084 + 3x = 0.2$$
$$3x = 0.2 - 0.084$$
$$3x = 0.116$$
$$x \approx 0.03866667$$

Convert to a percent and round to two decimal places: $x = 3.87\%$.

Glossary

abscissa   Fancy-pants word for the x part of a coordinate pair.

absolute value   The nonnegative value of the number or expression.

area   The amount of space covered by a two-dimensional object.

associative property   The manner in which a sum or product is grouped has no affect on its value.

asymptote   Boundary line that a graph gets infinitely close to but never actually touches.

axiom   See *property*.

base (of an exponential expression)   In the expression a^b, the base is a.

bomb method   Technique used to factor a quadratic polynomial with a leading coefficient not equal to 1; also called *factoring by decomposition*.

coefficient   The number appearing at the beginning of a monomial; the coefficient of $12xy^2$ is 12.

common denominator   A denominator shared by one or more fractions; it must be present in order to add or subtract those fractions.

commutative property   The order in which you add or multiply values has no affect on the sum or product.

completing the square Process used to solve quadratic equations by manufacturing binomial perfect squares.

complex fraction A fraction that contains another fraction in its numerator, denominator, or both; also called a *compound fraction*.

complex number Has form $a + bi$, where a and b are real numbers; imaginary and real numbers are complex numbers as well.

composite A number that has factors other than the number itself and 1.

composition of functions The process of inputting one function into another, sometimes denoted with a circle operator: $(f \circ g)(x) = f(g(x))$.

compound fraction See *complex fraction*.

compound inequality One inequality statement that is comprised of two others, such as $a < x < b$.

compound interest A method of earning interest on the entire balance, rather than just the initial investment.

conjugate The quantity $a - bi$ associated with every complex number $a + bi$; the only difference between a complex number and its conjugate is the sign immediately preceding its imaginary part, bi.

constant A number with no variable attached to it.

constant of proportionality Real number that describes either direct or indirect variation.

coordinate pair The point (x,y) used to describe a location in the coordinate plane.

coordinate plane Grid used to visualize mathematical graphs.

Cramer's Rule Technique used to solve systems of equations using determinants of matrices.

critical numbers The values of x for which an expression equals zero or is undefined.

cross multiplication Method of solving proportions in which you multiply the numerator of one fraction by the denominator of the other and set those products equal.

cube root A radical with index 3.

degree The largest exponent in a polynomial.

denominator   The bottom number in a fraction.

dependent   Describes a system of equations with an infinite number of solutions.

determinant (of a matrix)   Real number value defined for square matrices.

direct variation   Exhibited when two values, x and y, have the property $y = k \cdot x$, where k is a real number.

discriminant   The expression $b^2 - 4ac$, used to determine how many real solutions a quadratic equation has.

dividend   The quantity b in the division problem $a\overline{)b}$ is the dividend.

divisible   If a is divisible by b, then $\frac{a}{b}$ is an integer—the remainder is zero.

divisor   The quantity a in the division problem $a\overline{)b}$ is the divisor.

domain   The set of real numbers that can be substituted into a function.

double root   Solution of a polynomial equation that occurs twice, the result of a repeated factor in the polynomial.

elements   The numbers, variables, or expressions within a matrix; also called *entries*.

equation   A mathematical sentence including an equal sign.

even   Describes a number that is divisible by 2.

exponent   In the expression a^b, the exponent is b.

expression   Mathematical statement that doesn't contain an equal sign.

factor   If a is a factor of b, then b is divisible by a.

fraction   Ratio of two integers.

Fundamental Theorem of Algebra   Guarantees that a polynomial of degree n, if set equal to 0, will have exactly n roots.

function   A relation whose inputs each have a single, corresponding output.

greatest common factor (GCF)   The largest factor of two or more numbers or terms.

grouping symbols   Symbols such as parentheses and brackets that separate the terms of an expression into one or more groups.

horizontal line test   Determines whether or not a function is one-to-one based in its graph.

i   The imaginary value $\sqrt{-1}$.

identity element   The number (0 for addition, 1 for multiplication) that leaves a number's value unchanged when the corresponding operation is applied.

imaginary number   Has form *bi*, where *b* is a real number and $i = \sqrt{-1}$.

improper fraction   A fraction whose numerator is greater than its denominator.

inconsistent   Describes a system of equations that has no solution.

index   In the radical expression $\sqrt[a]{b}$, *a* is the index.

indirect variation   Exhibited by two quantities, *x* and *y*, when the product *xy* remains constant: *xy* = *k*; also called *inverse variation*.

inequality   A statement with sides that are either unequal (when the symbol is < or >) or possibly unequal (if the symbol is ≤ or ≥).

integer   A number with no explicit fraction or decimal.

intercept   Point on the *x*- or *y*-axis through which a graph passes.

interval   Segment of the number line.

inverse functions   Functions that cancel each other out when composed with one another: $(f \circ g)(x) = (g \circ f)(x) = x$.

inverse variation   See *indirect variation*.

irrational number   A number that cannot be expressed as a fraction; its decimal form neither repeats nor terminates.

leading coefficient   The coefficient of the first term of a polynomial in standard form; its term contains the variable raised to the highest power.

least common denominator   The smallest possible common denominator for a group of fractions.

like radicals   Radical expressions that contain matching radicands and indices.

like terms   Terms containing variables that match exactly.

linear equation   An equation of the form *ax* + *by* = *c*; its graph is a line in the coordinate plane.

matrix   A group of numbers or expressions arranged in orderly rows and columns and surrounded by brackets.

mixed number   One way to express an improper fraction; it has an integer and fraction part written together, such as $5\frac{1}{2}$.

natural number   A number in the set $\{1, 2, 3, 4, 5, \cdots\}$.

negative   Describes a number less than zero.

number line   Graphing system with only one axis, used to visualize inequalities containing one unique variable.

numerator   The top number in a fraction.

odd   Describes a number that is not divisible by 2.

one-to-one   Describes a function when each of its inputs have a *unique* output (no inputs share the same output value).

opposite   The opposite of the real number n is $-n$, the number multiplied by -1.

order   Describes the dimensions of a matrix; a matrix with m rows and n columns has order $m \cdot n$.

ordinate   Fancy word for the y part of a coordinate pair.

parallel   Describes nonintersecting lines in the same plane; parallel lines have the same slope.

perfect cube   Generated by multiplying a value times itself twice; b^3 is a perfect cube because $b \cdot b \cdot b = b^3$.

perfect square   Generated by multiplying a value times itself; a^2 is a perfect square because $a \cdot a = a^2$.

perimeter   Distance around a two-dimensional object (the sum of the lengths of its sides).

perpendicular   Describes lines in the same plane that intersect at right (90-degree) angles; the slopes of perpendicular lines are opposite reciprocals.

piecewise-defined function   Made up of two or more functions that are restricted according to input; each component function that comprises the piecewise-defined function is only valid for specific x-values.

point-slope formula   A line with slope m that passes through the point (x_1, y_1) has equation $y - y_1 = m(x - x_1)$.

polynomial   The sum of distinct terms, each of which consists of a number, one or more variables raised to an exponent, or both.

positive   Describes a number that's greater than zero.

power   See *exponent*.

prime   A number or polynomial divisible only by the number itself and 1.

principal   The amount of money initially deposited into a bank account, usually discussed when calculating interest.

product   The product of a and b is $a \cdot b$.

property   A mathematical fact that is obviously true and accepted without proof.

proportion   Equation that sets two fractions equal, such as $\frac{a}{b} = \frac{c}{d}$.

quotient   The quotient of a and b is $a \div b$.

radical   The $\sqrt{}$ symbol.

radicand   In the radical expression $\sqrt[a]{b}$, b is the radicand.

range   All of the real numbers that are valid outputs of a function.

rational number   A number that can be written as a fraction, a terminating decimal, or a repeating decimal.

rational root test   Generates a list of all possible rational roots for an equation.

rationalizing the denominator   The process of removing all radical expressions from the denominator of a fraction.

real number   Any number (whether rational or irrational, positive or negative) that can be expressed as a decimal.

reciprocal   The reciprocal of the nonzero rational number a is $\frac{1}{a}$; the reciprocal of $\frac{c}{d}$ is $\frac{d}{c}$.

relation   A rule that pairs inputs with outputs.

root (of an equation)   See *solution*.

scalar   When a matrix is multiplied by a real number, that number is called a scalar.

slope   Number describing how "slanty" a line is; it's equal to the line's vertical change divided by its horizontal change.

slope-intercept form   The linear equation $y = mx + b$, where m is the slope and b is the y-intercept of the line.

solution   Values that, if substituted for the variable(s) of an equation, make that equation true.

square matrix   A matrix with an equal number of rows and columns.

square root   A radical with index 2.

standard form (of a line)   Requires that a linear equation have form $ax + by = c$, where a, b, and c are integers and a is nonnegative.

substitution   Replacing a variable or expression with an equivalent value or expression.

surface area   The amount of "skin" needed to cover a three-dimensional object, neglecting its thickness.

symmetric property   If the sides of an equation are reversed, the value of the equation is unchanged; in other words, if $a = b$ then $b = a$.

synthetic division   Technique used to calculate polynomial quotients when the divisor is a linear binomial.

system of equations   A group of equations; you are usually asked to find the coordinate pair or pairs which represent the solution or solutions common to all the equations in the system.

terminating decimal   A decimal that is not infinitely long.

terms   The clumps of numbers and/or variables that make up a polynomial.

test point   A value or coordinate used to determine which of the intervals of a linear inequality or regions of a system of linear inequalities comprise the solution.

undefined   Describes a fraction when its denominator is zero but its numerator is not.

variable   Letter used to represent an unknown or changing value.

vertex (of an absolute value graph)   Sharp point at which the graph changes direction.

vertical line test   Determines whether or not a relation is a function based on its graph.

volume   The amount of three-dimensional space inside an object.

whole number   A number in the set $\{0,1,2,3,4, ...\}$.

x-axis   Horizontal line on the coordinate plane with equation $y = 0$.

*y***-axis**   Vertical line on the coordinate plane with equation $x = 0$.

zero product property   If the product of two or more quantities equals zero, then one of those quantities equals zero.

zeros (of a function)   The x-values at which $f(x)$ equals zero; the zeros of $f(x)$ are also its x-intercepts.

Index

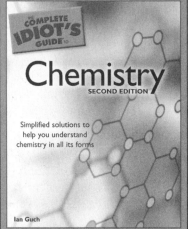
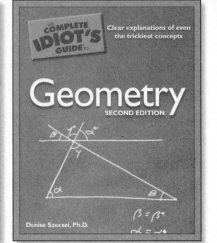

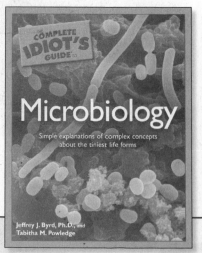
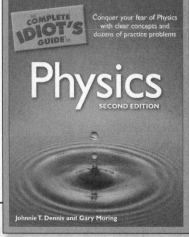